W9-DIU-147

St. Louis Community
College

Understanding Solid State Electronics

Written by: William E. Hafford
Staff, Texas Instruments Information Publishing Center

Eugene W. McWhorter
Longview, Texas
Staff, Texas Instruments Information Publishing Center

With Contributions by: Gerald Luecke, MSEE
Mgr. Technical Product Development
Texas Instruments Information Publishing Center

Ben Korte, Editor
Texas Instruments Information Publishing Center

P.O. BOX 225012, MS-54 • DALLAS, TEXAS 75265

This book was developed by:
The Staff of the Texas Instruments Information Publishing Center
P.O. Box 225012, MS-54
Dallas, Texas 75265

This book is produced as part of a TI Learning Center project on basic electronics, with principal contributions, in addition to the authors, as follows:

First and Second Editions — A.M. Bond, J.S. Campbell, J.R. Carter, Jr., J.E. Chambers, D.L. Garza, D.K. Gobin, G. Keegan, A.L. Feris, W.L. Kenton, L.J. LeVieux, G.P. McKay, J.R. Miller, E.G. Morrett, R.E. Sawyer, D.C. Scharringhausen, F.H. Walters, D.C. Ward, and others.

3rd Edition — Dr. D.L. Cannon, G. Luecke.

4th Edition — B. Korte, G. Luecke.

Word Processing:
Betty Brown

Design and artwork by:
Plunk Design

ISBN 0-672-27012-9
Library of Congress Catalog Number: 84-51250

Fourth Edition
Second Printing

About the cover:

Represented on the cover are the main topics of this book — semiconductor diodes (silicon, encased in glass), semiconductor transistors (silicon, encased in plastic), resistors, and DIP switches. All are used in solid state electronic circuits. Advances in techniques provide the means whereby the complete schematic shown is fabricated as one piece of semiconductor material called an integrated circuit.

Table of Contents

Preface

This book was created for the reader who wants or needs to understand electronics, but can't devote years to the study. The basic challenge was to explain engineering concepts without using mathematics — so you won't find any math in this book beyond a little fourth-grade arithmetic. A second challenge was to teach technical concepts to non-technical people, some of whom would have trouble wiring a doorbell — so you'll find that this book begins at the beginning and explains every new idea and expression along the way.

Most such attempts to popularize science leave a rather superficial knowledge of the field with the reader, no matter how conscientious he or she is. But people who have completed this book reassure us that it has equipped them to "hold their own," even in technical conversations with electronic engineers.

Periodically, because of the rapid advances in this field, new material must be added to keep the book abreast. This was the case for the third edition. Chapter 12 was expanded to include new up-to-date material on MOS and large-scale integrated circuits including microprocessors and microcomputers. A totally new Chapter 13 was added to cover integrated circuit techniques used for linear circuits. For the fourth edition, a new size, a totally new format with two-color printing, color enhancement of illustrations, and concise marginal statements for summary and review, contribute to a completely new presentation for the book.

We will have achieved our objective in publishing this work if it helps you to perform more effectively on the job, or helps you enjoy your hobby more. But we will have achieved a still more important objective if we succeed in heightening your awareness of the technology which, more than any other, is shaping the future of humankind.

1 WHAT ELECTRICITY DOES IN EVERY ELECTRICAL SYSTEM

Key Words

Alternating Current
Amplitude Modulation
Analog
Current
Digital
Direct Current
Electrons
Frequency
Frequency Modulation
Resistance
Voltage

Definitions are found in the glossary in the back of the book.

What Electricity Does in Every Electrical System

The function of every electrical system is either to process information, or to perform work, or both.

Let's jump right into our study of semiconductors and electrical systems, with two general statements that provide a starting point in simplifying the concepts. The first statement is: *All electrical systems either manipulate information, or do work, or do both. Regardless of the actual complexity of the system, everything that the system does will fall into one of these two categories — information or work.*

The second statement is: *All electrical systems are organized in a similar fashion.* We might term this "the Universal System Organization." Any system can be broken down into three basic elements of organization: The elements of *sense, decide,* and *act.*

The overall organization of every electrical system has three elements: sense, decide, and act.

In Fig. 1.1, we have a block diagram of the "Universal System." The system must have inputs, as shown by the input arrows. Typically, this input information is non-electrical — as, for example, pushing the starter button in an automobile. Then, between the boxes, we must have a flow of information, as shown by the arrows. Finally, in the act stage, we have a conversion of information into the desired action, as shown by the "action" arrows. This action can be either work or information in a desired form. Work, for example, might be the rapid turning of the bit by the motor in an electric drill; information in a desired form might be the numbers displayed as the answer on an electronic desk calculator.

All that our Universal System does — and all that any electrical system will do — is manipulate information or perform

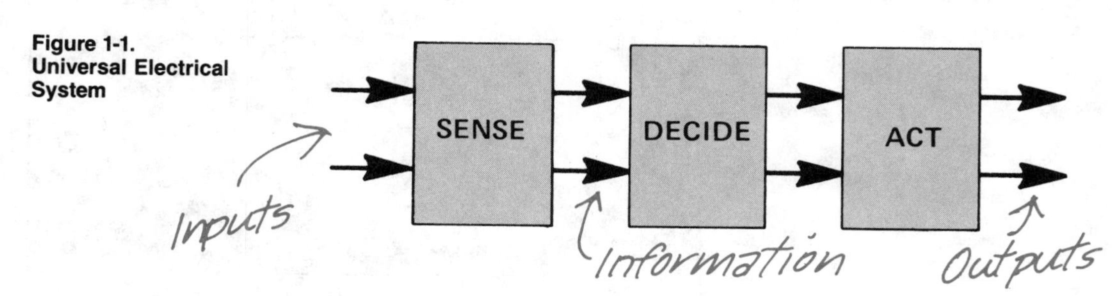

Figure 1-1. Universal Electrical System

work. And in every system, we will find information being input, we will find a flow of information internally, and we will find resultant actions. There's a familiar human analogy to this. You touch a hot stove. Your fingers *sense* the heat; this is the input of information. The information travels to your brain; this is the *decide* portion of the system. A decision is made, and the resultant information then travels to the *act* stage, your arm. At this point, the information is converted into the desired action, the quick removal of your hand; this removal of the hand is work. Alternatively, suppose your hand is stuck to the stove. Being unable to remove it, you call for help. This call may be considered a desired form of information, as opposed to work. So we see that our human system, like electrical systems, can be divided into the stages of sense, decide, and act.

A thermostatic heater control combines the three system elements—a temperature sense, a decision to open or close the gas valve, and an act of actually moving the valve—in a straightforward way.

Now, let's move a little closer to electronics. Figure 1.2 presents a functional or block diagram of a simple yet typical system, a thermostatic control system for a central heating unit. This system must have a temperature-sensing device, and it must have a control that can be set at the desired temperature. Both of these devices convert external information into a form that can be handled internally. The temperature-sensing device, a kind of thermometer, tells the system when the room temperature has fallen below, or risen above, the desired level. The control tells the system what that desired level is. Thus, these devices convert external information to internal information that can be handled

Figure 1-2. Thermostatic Control System

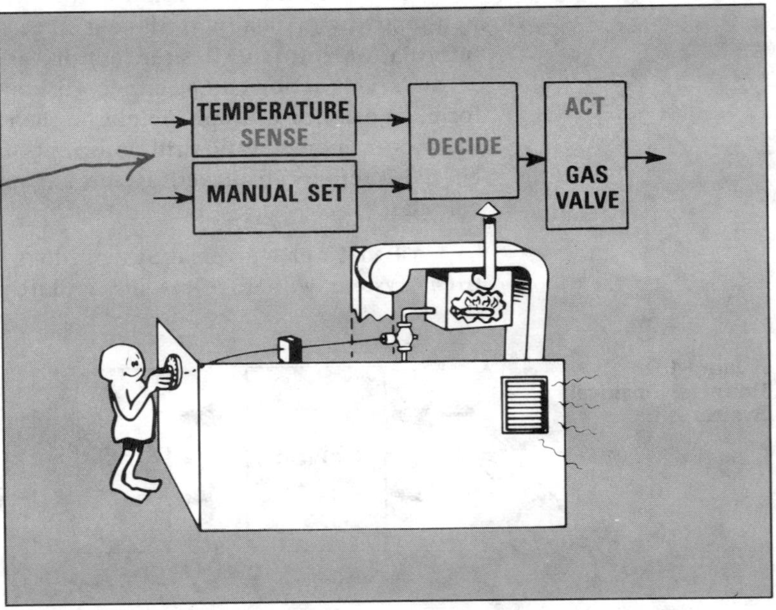

Information inputs- temperature and dial setting.

by the system. Next, the system has to use these two streams of information and arrive at a decision. In this case, it is a decision that in essence tells the fuel valve either to turn-on or to turn-off. If the decision is "turn-on," the fuel valve actuator converts this information into the action of moving the rather heavy valve parts. So here again we see how our system is organized in the universal manner: *Sense, decide, act.* And we dealt with either information or work: Information at the input, work at the output.

A phonograph "senses" input from the needle and the control settings, "decides" in the preamplifier how to interpret the information, and "acts" upon its decisions as it produces sound.

Let's take still another example. Figure 1.3 shows a block diagram of a phonograph or hi-fidelity system. First, we have input from the needle and cartridge; they perform a sensing function as the groove in the record passes under the needle. We also have input from the hand controls for volume and tone. The internal electrical information from these input devices usually goes to a preamplifier which, in effect, decides what the loudspeaker should do — the decision being based on the input information signal to the power amplifier. The amplified signal passes on to the loudspeaker, which finally acts to produce sound in the air. So, once more, we can divide the entire system into three segments: One that senses, one that decides, and one that acts.

**Figure 1-3.
Phonograph System**

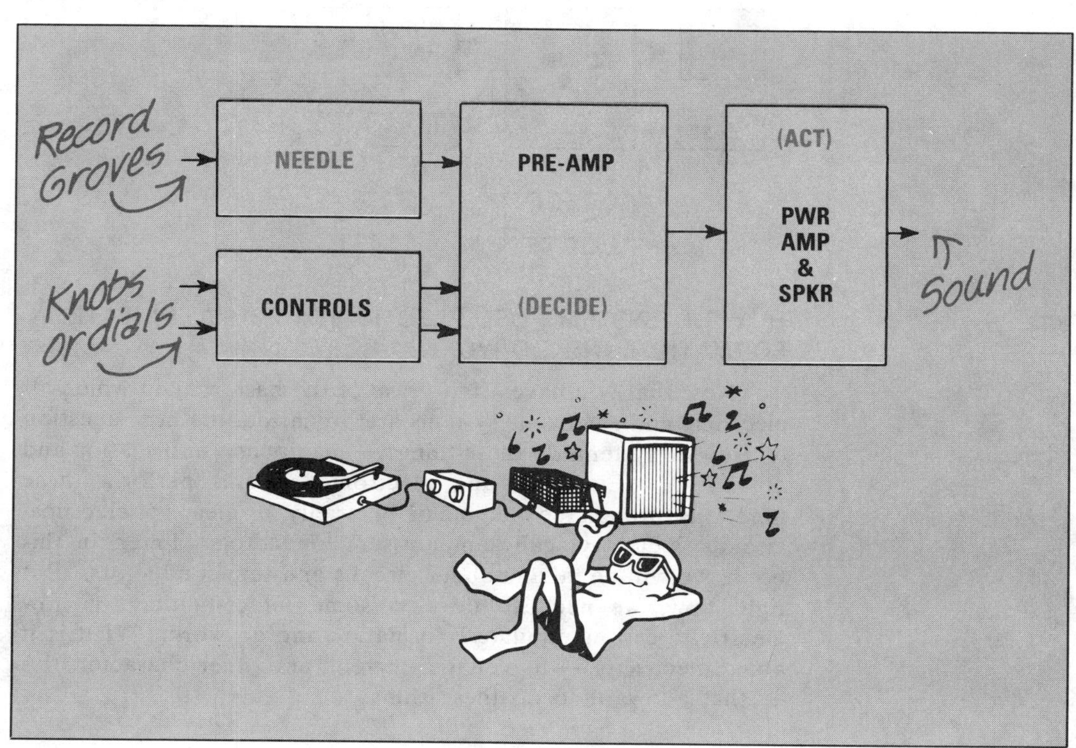

In electronic systems the elements of sense, decide, and act are commonly described as input, processing, and output.

You will hear other terms for these three system stages. We use the terms "sense, decide, and act" because they are graphic and easy to visualize, but the terms "input, process, and output" are synonymous with them. You will also hear the terms "input interface" and "output interface." These are very appropriate terms for the sense and act stages, since these stages do act as interfaces or "go-betweens," converting information and work between the external world and the electrical system.

The furnace control and the phonograph are relatively simple systems. But let's look at a computer, and see how this same Universal Organization exists in more complex systems. Figure 1.4 shows that the computer, too, is divided into the three typical segments. But here, we begin to indicate the added complexity of the computer by showing that the sense or "input" block is divided into two sections, accommodating two streams of information. Similarly, the act or "output" block is divided into two parts. The decide segment, in Universal Organization terms, is the central processing unit of the computer.

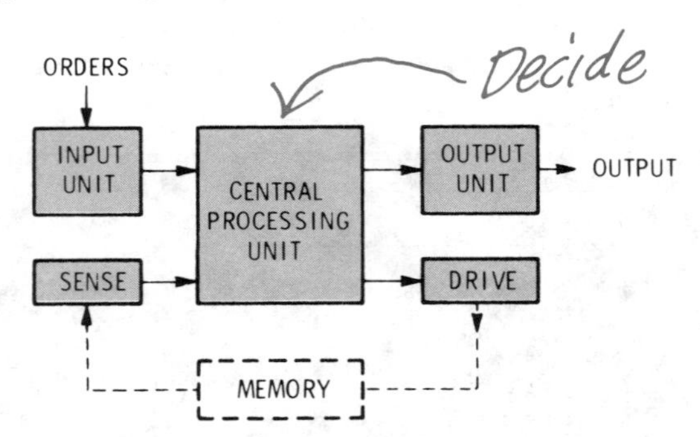

Figure 1-4. Computer System

HOW DO SYSTEMS USE ELECTRICITY TO MANIPULATE INFORMATION AND DO WORK?

Now that you have a fair grasp of the basic way in which all electrical and electronic systems are organized, the next question is "How do systems *do* these things — manipulate information and do work?" Electrical and electronic systems perform these functions through the medium of electricity, by means of electrical circuits, which typically employ semiconductors. Later, in this book, we will be talking about circuits and semiconductors. But right now, we need to develop some understanding of how electricity can manipulate information and do work. What is it about electricity — its voltage, current, and other characteristics — that allows it to do these things?

Electricity behaves much like water.

Electricity is really rather simple, because it behaves like a liquid. It flows like water. And like water, it tends to fill every space available to it. Electricity is made up of minute particles called electrons, which exist in every kind of matter. In a metal wire, electrons can be pumped like water by a generator or battery. Electrons repel each other, so they tend to reach the same density throughout a circuit, like water seeking the same level under the influence of gravity. Since there are so many basic similarities between the behavior of water and that of electricity, we can illustrate electrical characteristics by employing a water-flow analogy.

Figure 1.5 shows an open-top tank with water in it. We have a little man at a water pump who has filled the tank up to a certain depth by pumping water up from a lower level sluice of water. Water flows into the sluice at the point where another little man is measuring the depth of the water in the tank. The deeper the water in the tank, the greater the water pressure at the sluice. This pressure forces water to flow through the sluice at a rate of so many gallons per minute. If the water pressure at the sluice entrance is increased by raising the height of the water in the tank, the amount of water flowing through the sluice every minute will increase.

Electron pressure is voltage. Electron flow is current.

Electricity behaves in a similar manner. We have a source of electron pressure called a generator. The pressure on the electrons is measured in volts and may be produced by a battery, a piece of equipment called a power supply, a power generator plant, and so on. These perform the same function as our little man at the pump and the water tank. The flow of electrons is similar to the flow of water, except the electron flow rate is measured in amperes ("amps") instead of gallons/minute. Such electron flow is called current. Just as in the water system, increasing voltage (electron pressure) causes an increase in current (electron flow).

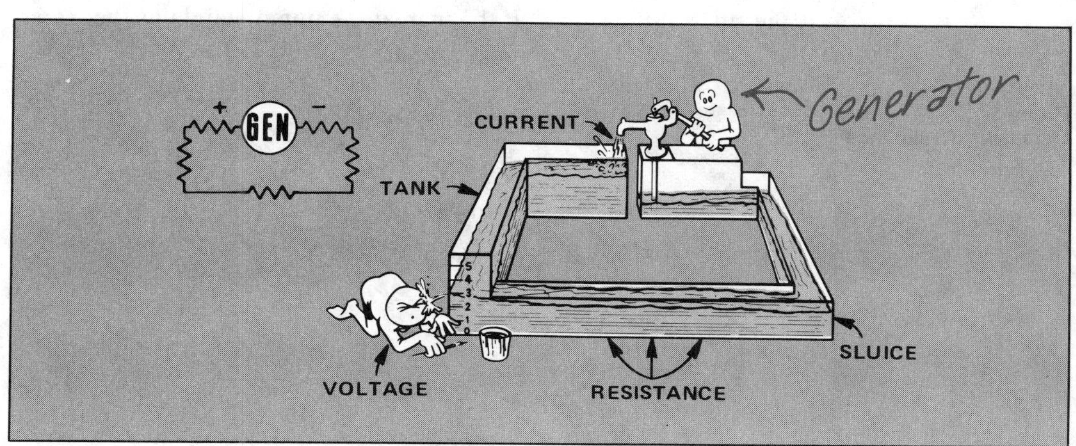

**Figure 1-5.
Water and Electricity Similarities**

Just as the rate of water flow depends on the water pressure and the capacity of the sluice, the amount of electrical current depends on the voltage and the resistance—opposition to current—of the conducting material.

The rate of flow of water can be changed by changing the dimensions of the sluice. More gallons/minute of water will be able to flow through a large sluice than through a sluice the size of a soda straw. The small sluice is said to resist flow more than the large sluice. Or we can say the large sluice conducts more water than does the small sluice. Electron flow can similarly be affected by the size and characteristics of the material it is flowing through. This effect is called resistance, and is measured in ohms. The resistance is defined as the ratio between electron pressure in volts and electron flow in amps as shown in Fig. 1.6. This figure also shows that flow increases with increasing voltage. The symbol commonly used for electron pressure is V (measured in volts). The symbol commonly used for electron flow is I (measured in amps). R is the symbol used for resistance.

Figure 1-6. Resistance—A Graph of Voltage Versus Current

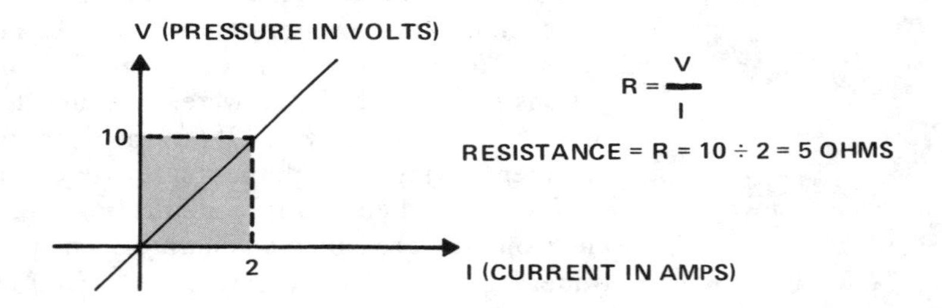

Resistance can be controlled by controlling the dimensions of the material in which the flow occurs. Figure 1.7 shows the sides of the sluice pulled in, constricting the passage, and thereby increasing resistance. We can do the same thing with the flow of electrical current by using a variable resistor. What happens when we pull in the sides of the sluice? If the man at the pump maintains the water

Figure 1-7. Increasing Resistance

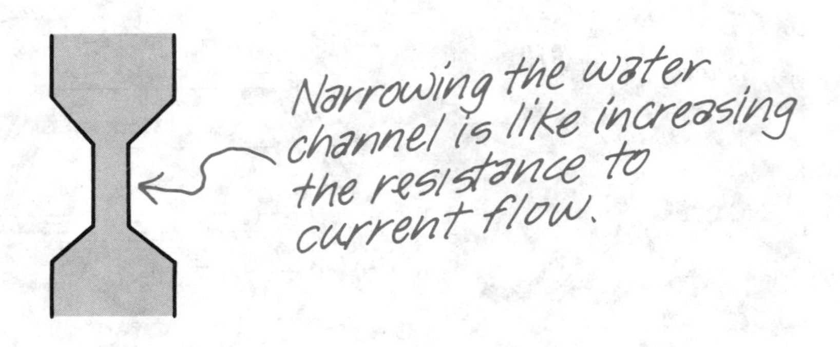

in the tank at the same height (pressure) the water flowing in the sluice will drop to a lower gallons per minute rate since less water can flow through the narrow channel per minute. The same thing happens with electricity; the voltage, current, and resistance are all related. If you change one, this changes one or both of the others.

A complete path for electrical current is called a circuit.

Electricity, like water, must flow in order to carry information or to perform work. To flow, it must come from somewhere and go somewhere. It's usually convenient to make it go in a circle to take care of this problem, and that's where we get the term "electrical circuit."

Before we move on, return for a moment to Fig. 1.5, and look at the schematic diagram of the electrical circuit represented by our water analogy. In the schematic, the circle represents the generator. The lines coming out of and going into the generator represent the conductors (or wires). And the zig-zag section indicates the resistance of the conductor; this zig-zag symbol also stands for a resistor.

HOW DOES ELECTRICITY CARRY POWER?

A circuit carries electrical energy from one point to another.

The useful thing about the flow of electricity is that it carries energy, or power, from one place to another. This energy can be put in at one point and used at another point. Figure 1.8 is a water analogy that illustrates this. Energy is put *into* electricity by pumping it from a low voltage to a high voltage. You get energy *out of* electricity by letting it fall from a high voltage to a low

**Figure 1-8.
Transferring Energy**

Increasing the voltage or current in a circuit increases its power.

voltage. In the water analogy, the energy is being converted to the useful work of sawing wood, by a waterwheel. We can put more power into the waterwheel by increasing the voltage difference (the height of the waterfall), or by increasing the current (the flow of water).

Electrically speaking, the pump we see here is representative of any device that puts energy into electricity. We'll continue to call it a generator, which is a device that converts mechanical energy into electrical energy. But the water pump equally well represents a microphone, which converts sound energy into electrical energy. The waterwheel represents any device that converts electrical energy back into external energy — for example, a motor, which produces mechanical energy, or a loudspeaker, which produces sound energy. For the sake of simplicity, we'll generally refer to the waterwheel as a motor. Figure 1.8 also shows the schematic electrical diagram equivalent to this simple circuit.

Energy that is not used for work produces heat.

Now let's see what happens if we remove the waterwheel from the waterfall as in Fig. 1.9. As far as the rest of the circuit is concerned, removing the wheel changes nothing. Water continues to flow across the voltage drop at the waterfall, and the only change is that no work is coming out. The waterfall is now simply the equivalent of a resistor. But what is happening to the energy — the work — that is still being put into the water by the pump? It is simply being wasted away by the friction, or resistance, in the waterfall. And like any friction, it produces heat; that's where the energy is going — into heating up the water and sluice. The same thing happens in our electrical circuit. Any working device, like a motor, can be replaced in a circuit by a resistor, without having

Figure 1-9. Energy Wasted as Heat

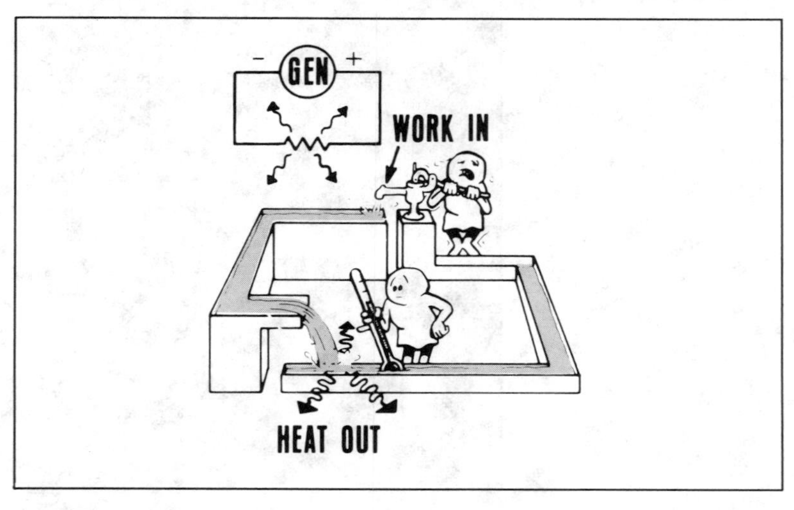

any effect on the circuit except that the work that might be done is instead wasted as heat, and the resistor gets hot. This is the way electrical heating elements and light bulb filaments work.

The point here is that any time electricity flows from a higher to a lower voltage — whether it's just from one end of a wire to the other, or through a resistor, or through a motor, or any other device — energy has to come out of it. If you don't recover the energy as some sort of work, it simply heats up the wire or the device. In the schematic of Fig. 1.9, we see this heat, represented by arrows, radiating from the resistor.

HOW DOES ALTERNATING CURRENT DIFFER FROM DIRECT CURRENT?

Direct current flows in a single direction. Alternating current switches directions continually.

In the circuits we have seen so far, the current flows in one direction. This is called "direct current," or "dc." An alternating-current circuit works just like a direct-current one, except that a special generator is required to pump current first in one direction through the circuit and motor, and then in the other direction. And a special motor is used, to recover work from current going in either direction.

Figure 1.10 shows an alternating-current circuit in hydraulic terms. This circuit is equivalent to the direct-current motor and generator circuit we saw in Fig. 1.8. The special pump represents

Figure 1-10. Alternating-Current Analogy

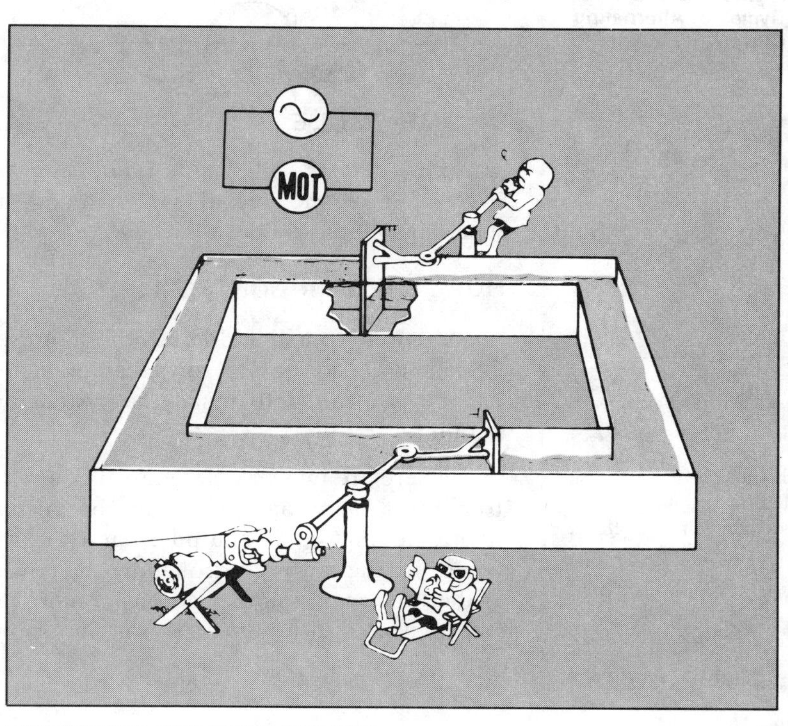

an ac generator. The paddle or piston connected to the pumping lever pushes water first in one direction and then the other. This produces a higher voltage first on one side of the paddle and then the other, so that the current around the circuit and through the motor alternates in direction. This pump, like the dc kind, simply puts energy into electricity.

The ac motor is represented by another piston-like paddle on a lever, just like the generator. When the voltage is higher on the left than on the right side of the paddle, it moves right, allowing some current to flow to the right; then the generator makes the voltage on the right side higher, and the paddle and current moves left. The water does work during each stroke and the work in this case goes into sawing wood.

WHAT IS ELECTRICAL FREQUENCY?

The rate at which alternating current changes direction is measured in cycles per second (hertz).

The frequency of alternating current is just the measure of how often it changes direction. That is, how many times every second a current goes through a complete "cycle," turning around backward and then going forward again. One cycle per second is called one "hertz." See Fig. 1.11. Of course, real electrical circuits use much higher frequencies than would be possible with our

Figure 1-11. Cycles of Alternating Current

ONE CYCLE

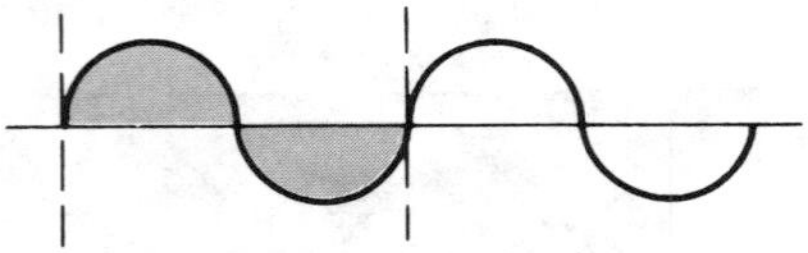

water model. You'll hear of kilohertz, meaning thousands of cycles per second, megahertz, meaning millions, and gigahertz, meaning billions.

HOW IS POWER CONTROLLED?

By now, you should have a basic grasp of the way electricity flows and carries power, so we can go a step further. This power can be controlled to make the system perform in the desired manner.

Either power can be varied continuously or simply switched on and off.

There are two ways to control power. The first way is simply to control the amount of power you put into the circuit. In our hydraulic analogy of the pump and the waterwheel, the power going into the saw is controlled by the power going into the pump. If the little man pumps vigorously, more power comes out of the waterwheel. If he slows down, less power comes out of the waterwheel. Typically, however, the power available to electrical systems is uncontrolled at its source.

The second way to control power is at some point in the circuits other than the power source, and this is more common. Figure 1.12 illustrates how this can be done — note the little man with the sliding dam, or "weir." Suppose the man at the pump is working at a steady rate; how can we vary the cutting power of the saw? By sliding the weir in or out, the control man can choke off or open the flow of water through the channel. So the man at the sliding weir controls the power driving the saw. He can make the saw cut fast or slow; he can turn it on or turn it off.

The man at the dam is representative of all that can be done to the flow of electricity within a circuit between the power source and the point of use. It can be throttled, or it can be switched on and off. This little concept is so important to remember when we are considering ways in which electricity can be used, that it bears repetition once more, in other words: *We can only do two things to electricity between the power source and the point of use. We can switch it — as in an "on or off" function — or we can regulate it, as when we vary the resistance.*

Figure 1-12. Power Control by Regulation

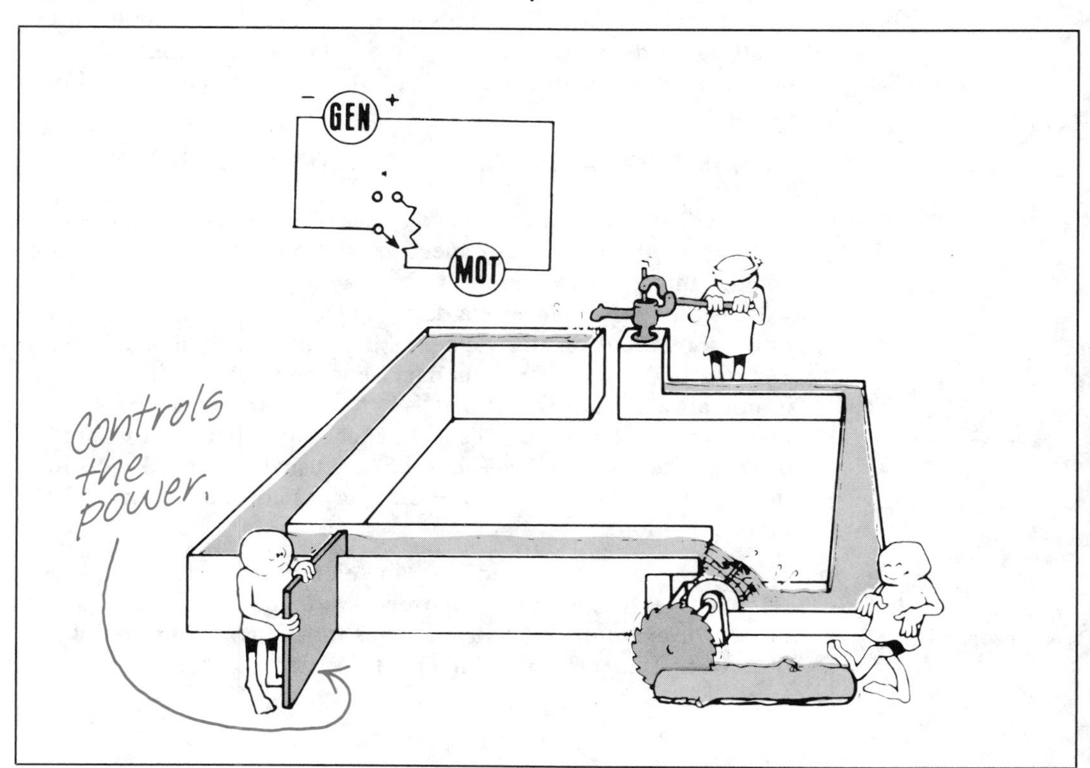

The schematic of Fig. 1.12 depicts what we have been talking about in electrical terms. It shows the generator (pump) and the motor (waterwheel). Between them in the circuit is the variable resistor (weir); this variable resistor can also act as an on-off switch.

So far, we haven't said much about semiconductors, and you may be asking the question — what does all of this have to do with semiconductors? The answer is that some knowledge of the simple basic concepts will provide you with the essential framework — some ways of thinking, if you will — that will greatly simplify your understanding semiconductors. For example, once you realize that every system can be divided into sense, decide, and act segments, it is easy to understand and remember where the various types of semiconductors are likely to be used. You can understand why light sensors are found chiefly in the *sense* portion of a system. You can see why the *decide* segment contains mainly small-signal diodes and transistors and integrated circuits. And it is easy to see why power semiconductors are found chiefly in the *act* segments, as well as in circuits that supply power to the whole system. And given this necessary framework, you can see for example why information in a system is generally manipulated by small-signal devices, and why work is most often controlled by power semiconductors. But right now, we need to lay more of our foundation.

INFORMATION AND WORK IN ELECTRICAL SYSTEMS

We can cover *work* quickly, because we saw examples of work as we discussed basic electricity. Converting our hydraulic analogy into electrical terms, we can say that in electrical systems, work is the performance of a discernible task — an electric motor lifting something, an electric heater providing warmth, a light bulb illuminating a room. The boundary between work and information is not always sharply defined — after all, the same light bulb illuminating a digit on the display panel of an instrument would be dealing with *information* —but we can usually make the distinction by asking "what's the chief end-purpose of the action? Work or information?"

Electrical systems that perform work generally require more power than those that process information.

Work, as performed by an electrical system, involves relatively large amounts of power. A common light bulb in the home, for example, typically requires more than a hundred watts of power. Information segments of a system, on the other hand, typically require only a few milliwatts — a few thousandths of a watt. But even though manipulating information requires much less power than work does, it is still achieved in one of the two ways we mentioned earlier; electricity can be switched, or it can be regulated. Let's see how these two methods can be used to send information.

HOW IS INFORMATION SENT BY THE DIGITAL METHOD?

The components of digital systems operate by switching on and off.

The method that involves sending information by *switching* is called the "digital method." All modern *digital* computers use this method of transmitting information. In contrast, the method of sending information by *regulating* is called the "analog method." Radios, phonographs, and *analog* computers provide examples of information carried by the analog method.

Since the digital method is somewhat simpler to understand, we'll consider it first. Digital computers use the same basic transmission method as a simple telegraph circuit.

When the switch is off in this example, there is no current. When the switch is on, current flows.

Let's look at the logical basis of telegraph code, to see how we might use such a technique in a computer. Figure 1.13 is a schematic of a simple, old-fashioned telegraph circuit. The power supply is a battery, which pumps electrons to a higher voltage on one side of the circuit than the other. The simple switch in the schematic is the telegrapher's transmitter key. And we've used a simple buzzer as the receiver. In the schematic, the switch is in the off (or open) position. Since the voltage on both sides of the buzzer is the same, the receiver is silent. When we press the key, turning the switch on, the voltage on the switch side of the receiver goes high, increasing the current flow and causing the buzzer to operate. When the switch returns to the off position, the current flow stops and the buzzer becomes silent.

We can say, then, that it is a change in voltage in the wire that

Figure 1-13. Power Control by Switching

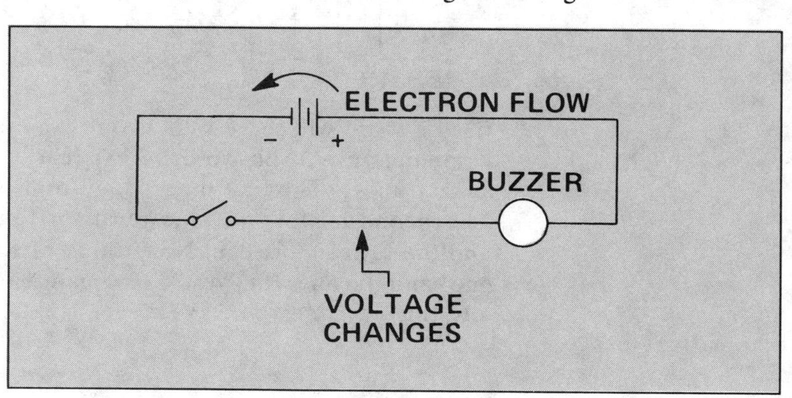

carries the information. We can visualize it as shown in Fig. 1.14. The level of the bottom horizontal lines represents zero voltage,

Figure 1-14. Dot and Dash as Digital Information

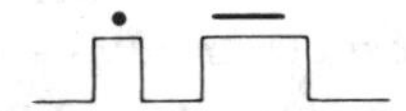

meaning the switch is off. When the switch is turned on, the voltage rises to the higher level indicated by the upper horizontal lines. If the switch is closed for a short time, we get a dot in Morse code. If it's held closed for a longer period, we get a dash. The curve shown gives us a dot-dash, which is an "A" in Morse code. That's all there is to it — switch on, switch off.

Now let's see how this digital method works in a computer. Digital computers are designed to handle numbers, not letters. But Morse code numbers are cumbersome, with (in International Code) five characters for each digit, so computers use a more efficient code called the "binary number code."

In digital systems, "off" and "on" or "low" and "high" are usually designated 0 and 1.

Here's how it works. We usually let a low voltage represent a zero; the higher voltage then represents a one. Figure 1.15 shows

Figure 1-15. Voltage Levels in a Digital System

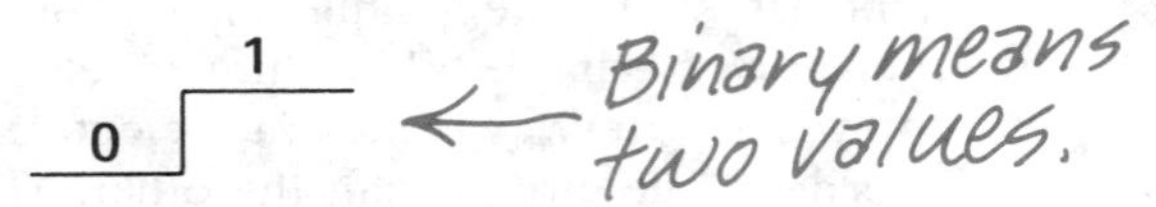

the voltage curve. Since all we can transmit in the binary number code is zeros and ones, how can we extract any intelligence from the code? Figure 1.16 shows a five-bit word; each zero or one is called a "bit," and a given number of bits makes up a word. This

Figure 1-16. Binary Code

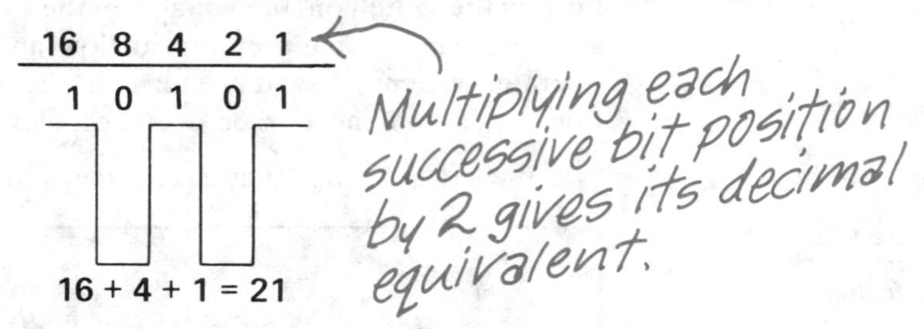

five-bit word will serve well as our example, even though typical computers use 32-bit words. Let's read this word as a number in binary code. The first bit reading from the right stands for one; the second bit for two; the third for four; the fourth for eight; and the fifth for sixteen. Now, think of the zeros as standing for "no" and the ones for "yes." So we can read the word from right to left, this way: Yes, we have a one. No, we don't have a two. Yes,

Individual 0's and 1's, called "bits," are grouped into "words" that form binary codes.

we have a four. No, we don't have an eight. Yes, we have a sixteen. Add up the values we do have, as we have done on the bottom line of the figure, and you get twenty-one. So twenty-one is the number represented by this word: 10101 in binary code.

It's easy to see how we can add more bits to the left. The next bit would represent thirty-two, the one beyond it sixty-four, beyond that one hundred twenty-eight, etc. In this way, we can send numbers as big as we want. And, of course, we can also encode decimal fractions. Digital computers use many other codes, such as binary-coded decimal, Gray code, and for letters, the Hollerith code. But all of these codes use just zeros and ones, so they're all binary codes. "Binary" means "two-state," on or off.

This simple principle of transmitting digital information has remained the same from the old-fashioned telegraph system through today's most modern and powerful digital computer. You've probably heard of Boolean algebra — it's an entire system of complex mathematics based on binary counting, that enables computers to perform highly sophisticated computations.

HOW IS INFORMATION SENT BY THE ANALOG METHOD?

In analog systems, electricity is a continuously varying quantity (not simply "on" or "off").

Since the only two ways we can control the flow of electricity are by switching it or regulating it, and the digital method uses switching, it follows that the only other method available must employ regulation. It does, and we call this the "analog method." To explain the analog method, we can use essentially the same circuit we employed in discussing the digital method. However, in Fig. 1.17, we have replaced the simple switch of the digital method with a variable resistor to regulate voltage. And instead of a buzzer, we use a meter to measure voltage. (This is a special meter, a galvanometer whose dial is calibrated in volts.) So now, the variable resistor regulates the voltage in the line to the meter.

Figure 1-17. Regulated Voltage as Analog Information

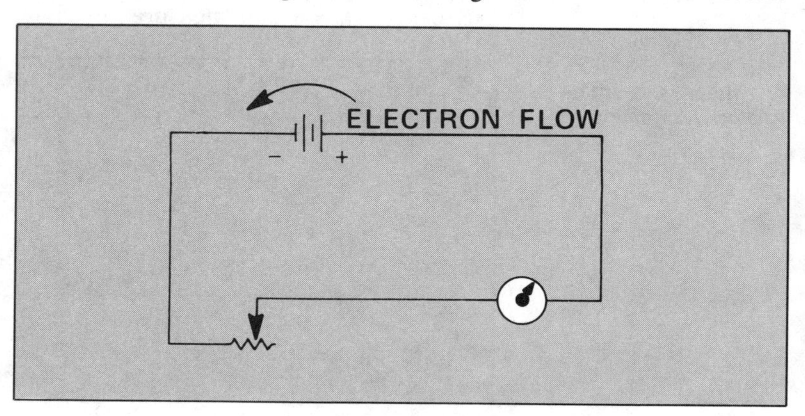

Analog systems represent information as actual levels of voltage or current, rather than as combinations of binary numbers.

In this analog method, we will let some measurement of the electricity in the line stand directly for the number we want to transmit. If, for example, we let a *voltage*-level measurement stand for the number, we have a *voltage* analog system. Suppose that we regulate the voltage to 10.5 volts, by using the variable resistor. Then, when we read the voltmeter, we would read the actual number 10.5. Or, by arranging the code with the receiver, it could mean double 10.5, or the square of 10.5, etc. If we changed the voltage by regulating the variable resistor, say to 2.36, we then transmit a different number.

A tremendous variety of electrical systems use voltage analog to transmit information. Most old-fashioned automobile fuel gauges work this way. A float in the gas tank controls the variable resistor. As the level of gas changes, the voltage going to the gas gauge changes. Such a gauge is really a voltmeter whose dial is marked from empty to full, instead of in volts. Other examples of voltage analog devices include analog computers, where voltage stands for numbers or mathematical functions of numbers. And in telephones, the voltage stands for fluctuating air pressure, which the ear interprets as sound.

Measurements other than voltage can be used to transmit information. *Current* analog systems, for example, operate the same way as voltage analog systems, except that they depend on measurements of current, or "amps," instead of voltage.

In amplitude modulation, the height of a radio wave is an analog representation of information.

An interesting variety of voltage analog is called "amplitude analog" — or more commonly, "amplitude modulation." In Fig. 1.18, we have replaced the battery of Fig. 1.17 with an alternating-current generator. Now, the voltmeter will oscillate constantly, as the generator produces alternating voltage, first high with current going in one direction, then low as the current goes in the other direction. Obviously, we can't establish a constant voltage level in this case, but we can measure the height, or amplitude, of the

Figure 1-18. Alternating Current as Analog Information

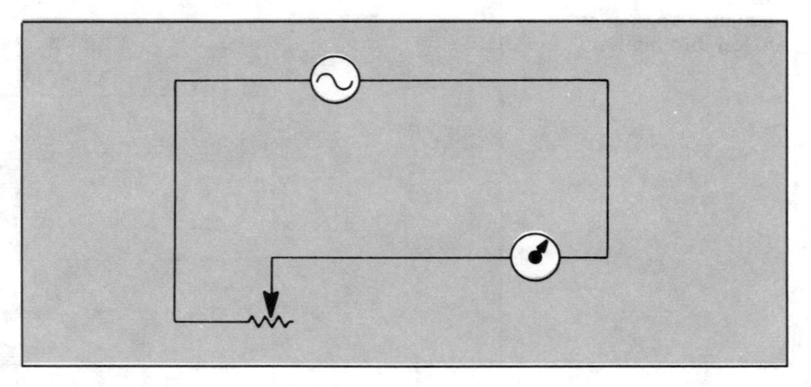

waves. By varying the resistance, we can change the height of the waves, as in Fig. 1.19. Thus, we can let the amplitudes stand for the

Figure 1-19.
AM Wave

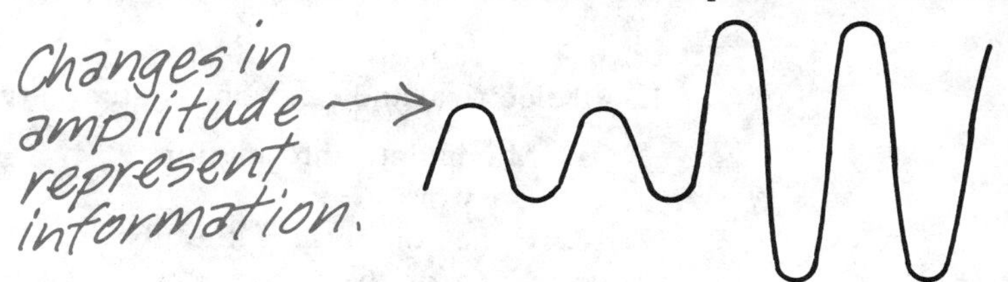

numbers we want to transmit. AM radios get their name from the fact that they operate on the principle of amplitude modulation.

In frequency modulation, the frequency of a radio wave is an analog representation of information.

Still another analog method is frequency modulation. This technique is employed in FM radios. It depends on waves just like AM. But instead of measuring the height of the waves, we measure their *frequency*. FM waves are shown in Fig. 1.20. Assume that the waves at the left end, which are close together,

Figure 1-20.
FM Wave

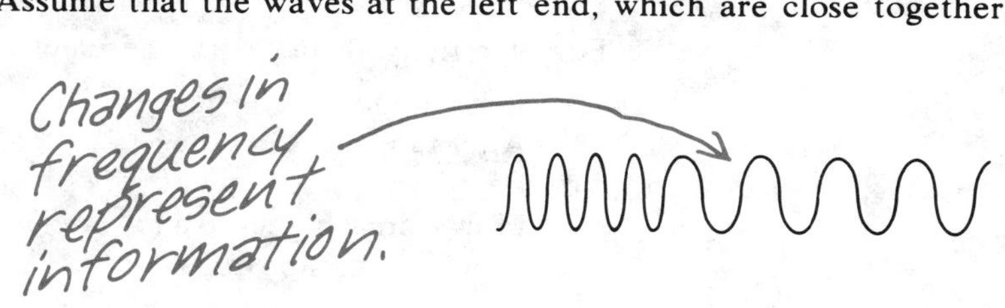

represent ten cycles per second (ten hertz). And assume that the waves near the right end, where they are farther apart, change to half that frequency, or five hertz. Now we have represented the numbers ten and five.

There are still other analog methods, of course, but the ones we've described are by far the most popular. In summary, we can say that all *analog* methods are based on *regulating* various properties of electricity. On the other hand, all *digital* methods are based on switching electricity *on and off.*

In the next chapter, we will see what goes on inside the three boxes of sense, decide, and act. Before going on, however, why not take the quiz for Chapter 1? You'll find the answers on the last pages of this book. By taking the quizzes that appear at the end of each chapter, you can assure yourself that you've mastered the information you need to know, to profit fully from later discussions.

Quiz for Chapter 1

1. All electrical and electronic systems are designed and built to:

- ☐ a. Manipulate information
- ☐ b. Do work
- ☐ c. Either or both of the above
- ☐ d. None of the above

2. The three basic elements of all electrical systems are:

- ☐ a. Sense, detail, act
- ☐ b. Sense, decide, act
- ☐ c. Input, act, output
- ☐ d. Sense, decode, act
- ☐ e. None of the above

3. For electricity to do work the electrons must:

- ☐ a. Be switched
- ☐ b. Alternate in direction
- ☐ c. Be of very good quality
- ☐ d. Flow from a higher to a lower electron voltage
- ☐ e. None of the above

4. Three of the most important factors controlling the flow of electrons in a circuit are:

- ☐ a. Voltage, current, and resistance
- ☐ b. Resistance, reactance, and current
- ☐ c. Voltage, current, and wattage
- ☐ d. Voltage, electromotive force, and current
- ☐ e. None of the above

5. Between the source of power and the point of use, all that can happen to electricity is:

- ☐ a. It can alternate in direction
- ☐ b. Its voltage can change
- ☐ c. It can manipulate information
- ☐ d. It can be switched and regulated
- ☐ e. None of the above

6. Analog and digital refer to the two ways electricity can:

- ☐ a. Be switched
- ☐ b. Be regulated
- ☐ c. Carry information
- ☐ d. Flow like water
- ☐ e. None of the above

7. In binary code, a "bit" means:

- ☐ a. 10
- ☐ b. 1
- ☐ c. 1 or 0
- ☐ d. 2
- ☐ e. None of the above

8. Wasted electrical energy in a circuit is dissipated as:

- ☐ a. Vibration
- ☐ b. Heat
- ☐ c. Excessive current flow
- ☐ d. Resistance
- ☐ e. None of the above

9. Increasing the resistance in a circuit causes the flow of electrons to:

- ☐ a. Be decreased
- ☐ b. Be stopped
- ☐ c. Be speeded up
- ☐ d. Remain constant
- ☐ e. None of the above

10. Frequency of alternating current is the rapidity with which the electrical current changes direction, and is measured in:

- ☐ a. Amperes
- ☐ b. Watts
- ☐ c. Hertz
- ☐ d. Ohms
- ☐ e. None of the above

2 BASIC CIRCUIT FUNCTIONS IN THE SYSTEM

Key Words

Amplifier
Base (P-region)
Collector (N-region)
Control Circuit
Emitter (N-region)
Modulator
Oscillator
Working Circuit

Definitions are found in the glossary in the back of the book.

Basic Circuit Functions in the System

Now that we have analyzed the common characteristics of all systems, we're ready to talk about circuits. Since we'll have to bounce back and forth a bit between various levels of organization within a system, let's get these levels clearly in mind. The highest level is the *system* (a radar system, a TV set, a clock, a radio); within each system are three *stages* (sense, decide, act); within each stage are one or more *circuits* (tuning circuit, counting circuit, light-sensing circuit); within each circuit are one or more *components* (transistors, diodes, rectifiers, integrated circuits, resistors, capacitors). So a single system has only three stages, but may have thousands of circuits and millions of components.

Much of what we will discuss in this chapter is based on a single fact we learned in the last chapter: There are only two things that can be done to electricity between a power source and a point of use — it can be switched or it can be regulated.

In the first chapter, we discussed some elementary methods of switching and regulating. We saw that electrical power can be regulated by a variable resistor. One common type of variable resistor is the rheostat. By physically turning the knob on the rheostat, we can change the amount of resistance, and thus decrease or increase the brightness of a light, or control the volume of a radio. And we saw how we can use a hand-operated switch to send telegraph messages.

The invention of vacuum tubes made electronic circuits possible, because it provided a means of controlling power electrically, and at very high speed.

But it's quite obvious that manual switching and regulating are totally unsuitable for modern electronics. How could we possibly build a workable system if we required manual switching and regulating in thousands of different circuits? The answer, of course, is that we could not build a system of any sophistication. The great breakthrough that made modern electronics possible was the invention of the vacuum tube. It provided a method for controlling electrical power by electrical means, rather than by mechanical or manual methods. The great benefit of the vacuum tube was that it could perform these switching and regulating functions at high speed — millions of times per second.

Transistors can perform the switching and regulating functions of vacuum tubes in less space, with greater reliability, and with lower power consumption.

Invention of the transistor, in turn, provided major improvements over the vacuum tube. Today, it is the very heart of modern electronics. The transistor does the same things that a vacuum tube does: it switches and regulates by electrical means. Compared to the vacuum tube, however, the transistor has many outstanding advantages: It requires no heater current, it is very small and light, it is mechanically sturdy and long-lasting, it operates at desirably low voltages but can carry relatively high current, and it is thousands of times more reliable than the vacuum tube. We'll be examining the transistor and its applications in detail later in this book. But for now, we must consider the transistor's functions of switching and regulating as our jumping-off point.

WHAT MAKES EACH SYSTEM STAGE WORK?

Each system stage is made up of one or more circuits — circuits in varying types and numbers depending on the purpose and complexity of the system. These circuits, operating individually and collectively, enable the system to operate in the desired manner. And it is by virtue of switching and regulating that we make the circuits perform as required. To comprehend the circuits themselves, we must first find out how their basic component, the transistor, behaves and what it does in the circuit. The most general and basic circuit arrangement is shown in Fig. 2.1. You'll recognize it as the same circuit we used in Chapter 1.

It has a power source, a working device, and a control. The control, as we saw before, is a device that switches or regulates. Let's see now how a transistor can be used as a regulator in this

Figure 2-1.
Basic Electric Circuit

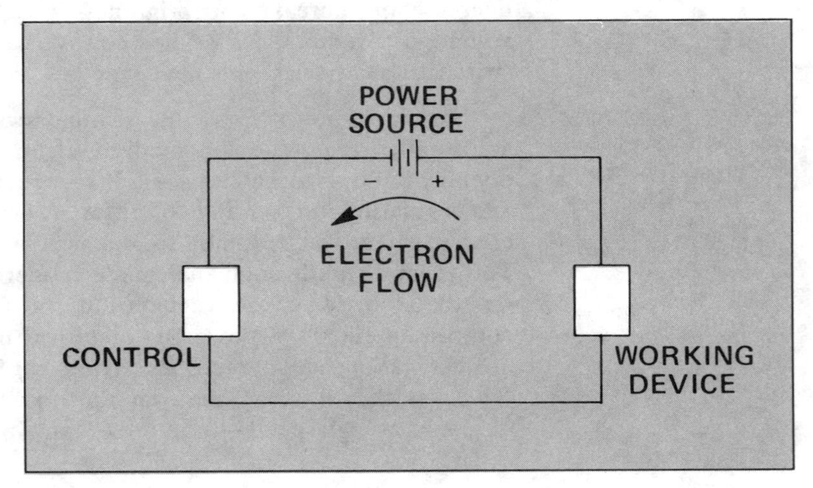

Circuits that regulate rather than switch are commonly called amplifiers.

same circuit. Since electronic engineers call regulating-type transistors "amplifiers," from this point on, we'll use the term "amplifier."

Before we put an amplifying transistor into this circuit, let's see how a transistor is constructed. The heart of every transistor is a small piece of semiconductor material, most often germanium or silicon. As the cross-section of a transistor in Fig. 2.2 shows, the

Figure 2-2.
Transistor Cross-Section

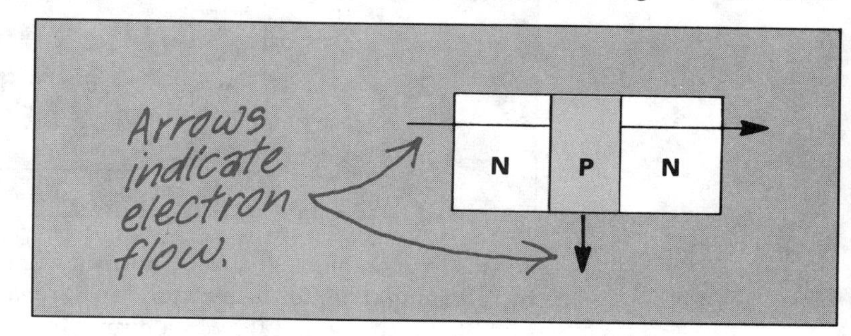

transistor is processed so it has three distinct sections, or regions, designated as either "P type" or "N type." Later on, we will cover what is meant by P and N.

This transistor is an NPN type. Later on, we'll also explain another type, the PNP. This piece of semiconductor material can be made to act as a variable resistor, or as a switch. It can be made to conduct current, to throttle it partially, or to block it entirely.

Let's see how this works by putting our transistor cross-section into our basic circuit schematic, as shown in Fig. 2.3. Our power source is still a battery, and for our point of use, we'll use a loudspeaker. We've connected a microphone to the P region of the transistor.

A transistor has three sections: emitter, collector, and base. The emitter emits electrons, the collector collects electrons, and the base controls electrons.

Before the microphone is connected in the circuit, nothing will happen. The transistor will merely block the flow of electricity from the battery to the loudspeaker. To make current flow, we must withdraw electrons from this central region, called the "base," to permit the current to flow from one N region to the other. The more electrons we withdraw, the more current will flow. One N region is called the "emitter," because as we extract electrons from the base, this region will *emit* electrons across the base region. The other N region is called the "collector," because it's the region where the flowing electrons will be *collected* and then pass on down the wire to the loudspeaker.

Current flows through an NPN transistor from emitter to collector when electrons are withdrawn from the base.

You'll notice that in Fig. 2.3 we have also hooked a wire from the microphone back to the emitter of the transistor. We had to

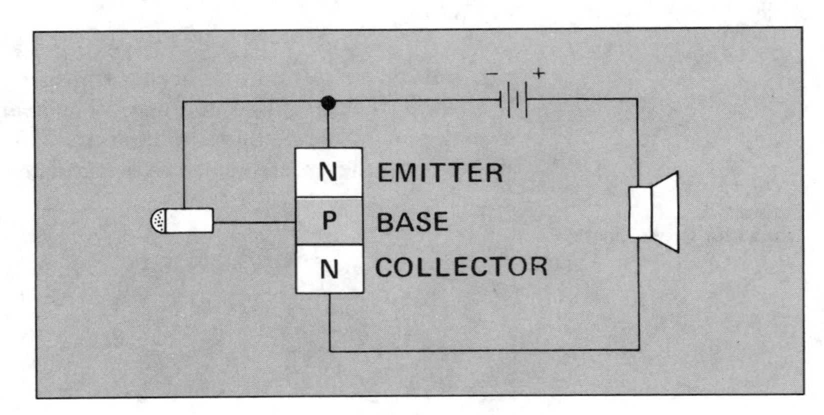

Figure 2-3. Transistor Cross-Section in a Basic Circuit

do this to give the electrons withdrawn from the base a place to go, by returning them to the emitter; you'll recall that in Chapter 1, we said electricity would flow in a circuit *only* if it had somewhere to come from and somewhere to go. This additional wire has completed what is called the "control circuit."

Now that you have the circuit concept firmly in mind, we might as well get professional and replace the cross-section drawing of the transistor with the proper symbol for a transistor, as in Fig. 2.4. In the symbol, the vertical line represents the base.

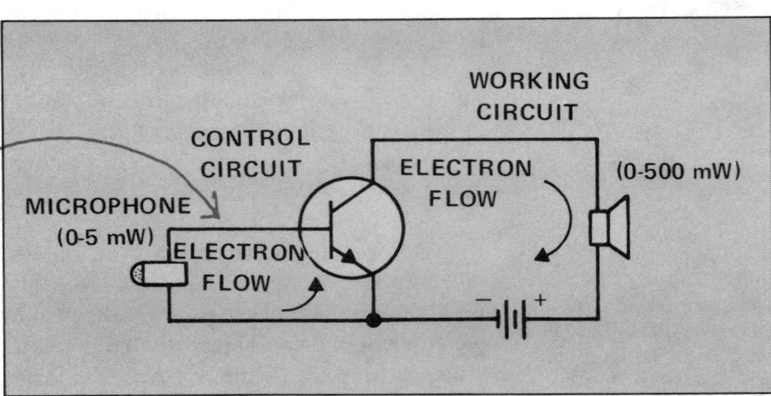

Figure 2-4. Transistor as an Amplifier

Electron flow is against the transistor arrow.

The plain diagonal represents the collector, and the diagonal with the arrow represents the emitter. The emitter arrow always points in a direction that is *opposite* to that of electron flow. The symbol is properly completed by drawing a circle around it, but the same symbol without a circle has the identical meaning.

HOW DOES A TRANSISTOR ACT AS AN AMPLIFIER?

We'll use Fig. 2.4 to demonstrate how an amplifier transistor works. We said that one of the marvelous qualities of a transistor is its ability to control electrical power by electrical means. In this example, the electrical control will be initiated by a microphone, a device that can produce fluctuating electric current corresponding to fluctuating sound waves. But the microphone can produce only a tiny trickle of power. If we attached it *directly* to a loudspeaker, you probably wouldn't hear any sound even with your ear pressed against the speaker. But with the simple circuit concept you see here, you can produce enough sound to keep the neighbors awake.

Just for sake of illustration, we've assigned the microphone a power output ranging from zero to five mW ("mW" equals milliwatt, which is one thousandth of a watt). But the power produced by the battery in the *main* circuit can range from zero to 500 milliwatts.

Small changes in the control circuit current are amplified in the working circuit.

Now let's assume that a single sound wave hits the microphone and creates a power output of three milliwatts. The microphone pumps a surge of electrons out of the base region into the emitter region. Now, as a result of the base current, a relatively large current will flow across the base region from emitter to collector and on down the line, through the coil of the loudspeaker. In this way, the current flow through the loudspeaker will be controlled, or amplified, in exact proportion to the much smaller microphone signal. The signal through the loudspeaker might be typically 300 milliwatts; this means that the three milliwatts of power produced by the microphone have been amplified one hundred times.

The amount of base current in the control circuit determines the amount of emitter-collector current in the working circuit.

Now suppose a second sound wave hits the microphone. This is a softer sound, and produces a power output of two milliwatts. Fewer electrons flow in the control circuit, so fewer are drawn from the base region, and this time the amount of power flowing through the transistor and the loudspeaker is only 200 milliwatts. Nevertheless, it has also been amplified one hundred times.

At all times, fairly precise proportions are maintained between the control circuit and the working circuit.

In other words, the power in the working circuit will always be an essential duplicate of the power in the control circuit — but much amplified. We can visualize the process as the voltage traces shown in Fig. 2.5. If the trace produced by the microphone looks like the tiny squiggle on the left, then the trace going to the

Figure 2-5. Amplification

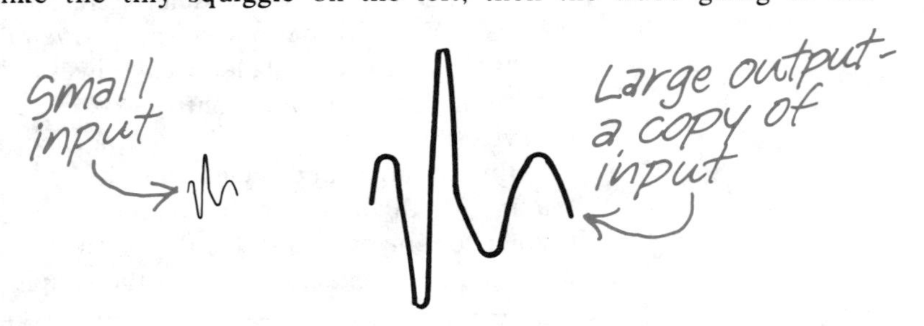

loudspeaker will look like the big squiggle on the right — a close copy of the little one but greatly amplified.

This is a good time to point out one of the characteristics of the transistor that makes it so useful in modern electronics. Sound waves fluctuate very rapidly, up to frequencies of about 30,000 cycles per second (30 kilohertz). The transistor has the capability of reacting to each of these fast fluctuations. In fact, high-frequency transistors can react *billions* of times per second.

HOW DOES A TRANSISTOR ACT AS A SWITCH?

Transistors can perform the function of a switch as well as amplify.

Now that we've seen the transistor in a regulating or amplifying function, let's look at it in a switching function. We'll use a telegraph circuit again as shown in Fig. 2.6. Once again, we have a battery as the power source, a buzzer as a point of use, and a transistor in the working circuit. Now, in the control circuit, we have a switch in place of the microphone. Since the switch cannot generate any power, we have a battery in the control circuit. The zig-zag resistors in the control circuit represent the resistance to the sound in sixty miles of wire. This resistance diminishes the control circuit power so much that not enough is left after sixty miles to actuate the buzzer. But the surviving power does provide enough energy to operate a transistor.

So when the transmitter key is pressed, a small current of electrons is withdrawn from the transistor base, a much greater

Figure 2-6. Transistor as a Switch

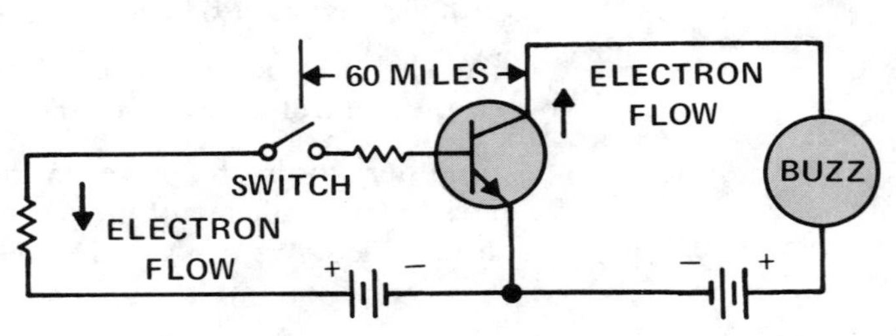

current flows in the working circuit, and the buzzer sounds. This transistor is acting as a switch in the working circuit.

This is a good time to point out an important fact about transistors: Every transistor has the capability of either *switching* the working-circuit current on and off, or *throttling* ("regulating") this current part-way between on and off. In other words, transistors are much like ordinary water faucets. You can turn them all the way off, all the way on, or only *partly* on. The only way an NPN transistor resembles an ideal switch more than an ideal amplifier is that it has a sort of "threshold voltage" of electron pressure that must be produced at the base by the control circuit before an appreciable amount of current will flow. (We will discuss this in Chapter 7.)

It is mainly the design of the control circuit that determines whether a transistor acts as a switch or an amplifier.

This brings up the question as to what determines whether a transistor in a particular circuit acts as a switch or as an amplifier. The answer is, "the control circuit mainly determines this." For example, compare the "control" portions of the amplifier circuit in Fig. 2.1 and the switching circuit in Fig. 2.6. In the amplifier circuit, the microphone (provided it achieves the "threshold" voltage) generates current fluctuating anywhere between zero and full power. But in the switching circuit, we have instead a battery and a switch. When the switch is off, the transistor is "off." And when the switch is on (provided the battery achieves the "threshold" voltage), the transistor is "on." This allows working current to flow in proportion to the steady, unvarying control current. (Ideally, we select a transistor that will be "all the way on" — called "saturated" — when this amount of current is flowing.) In other words, the microphone makes one transistor "regulate," and the transmitter key and battery make the other transistor "switch."

Although a transistor can either switch or regulate, normally it is manufactured to do one of these jobs better than the other. Consequently, you will hear some transistors referred to as "switches" and others as "amplifiers." For example, good amplifier transistors have a stable, moderate current gain (degree of amplification). But most switching applications require fast turn-on and turn-off speeds and low leakage. (These concepts will be discussed in Chapter 8.)

Since digital computer systems rely heavily on switching transistors, you should be able to envision the great value of the transistor in electronics.

HOW CAN I UNDERSTAND MORE COMPLICATED CIRCUITS?

Our discussion of the amplifier circuit and the switching circuit dealt with two very simple and crude circuits. If you were to build them, you'd find their performance disappointing. A single transistor was the only semiconductor component in each circuit, and you're aware that even such a simple system as a transistor radio involves at least six or seven transistors. Moreover, you know that a typical system includes not only transistors, but other components such as diodes, capacitors, resistors, inductors, and many more.

Our next step then, is to go back to the basic circuit and build it up to make it more complex and sophisticated — in other words, more typical. This time, we won't stop to explain any of the components other than the transistors. But don't worry if you can't understand all the circuitry as we go along. The chief object at this point is to gain an appreciation for some of the many ways semiconductors can work together in circuits to produce desired results.

The circuit of Fig. 2.7 is a little more complex than any we've seen. Yet you can recognize the working circuit on the right, and the control circuit on the left. We have labeled the power source "G" for generator. Actually, the power might be coming from a

Figure 2-7. Amplifier Circuit for Motor Control

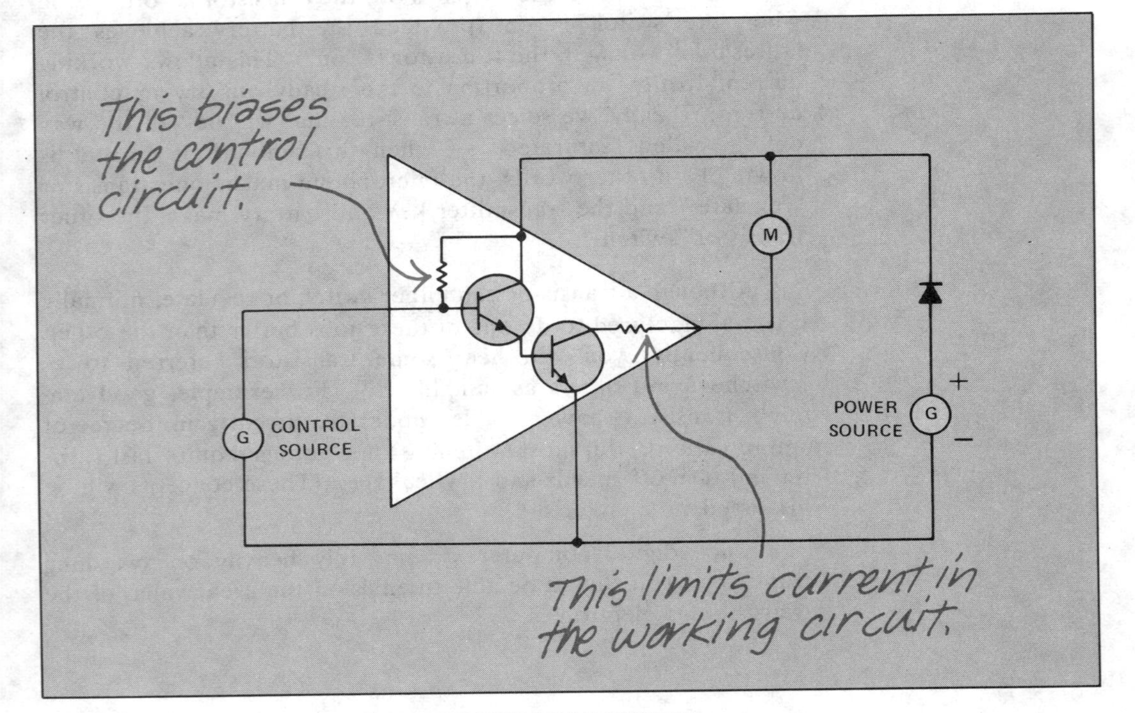

wall plug, or from solar cells in a satellite. A power source can be anything that pumps electrons. We also need a power source in the control circuit, and we have labeled that "G" also; but as you saw in an earlier example, it could be a microphone. We have labeled our working device at the point of use, "M" for motor.

"Biasing" the control circuit sets a steady-state level of current flow.

Suppose we want to regulate the speed of the motor — not from stop to full power, but from half power to full power. In other words, we want the motor to run at least at half power all the time, even if we are not pumping any electrons out of the base by using the generator in the control circuit. To accomplish this, we can add to the control circuit, and put in a resistor of a certain value; in this way, we create a "biasing circuit," and sufficient electrons will always be withdrawn from the base to keep the motor at half speed. Then when we pump from the control generator, we will be withdrawing even more electrons, so the motor speed will increase above half power.

Now, let's add an accessory component in the working circuit. Suppose the generator produces more current than the motor can handle without overheating. A resistor placed in the working circuit between the transistor and the motor — or anywhere in the working circuit — will limit current and prevent overheating in the motor. Let's further assume the generator to be an ac generator, but our motor is dc. So we've put a diode (rectifier) into the working circuit near the generator. This device allows electrons to flow in one direction only, changing the alternating current from the generator into direct current for the motor. The arrow in the rectifier symbol, incidentally, points *opposite* to the flow of electrons.

Using a low-power transistor in the control circuit of a high-power transistor lets a small input regulate a very large output.

And finally, suppose that our motor is very large. Our controlling transistor must then be a large, high-power type to handle the large current flow. But suppose that our control generator does not have the capacity to withdraw enough electrons from the base to operate the transistor. We simply use a smaller, *low-power* transistor that *can* be operated by the control generator first, and let this transistor, in turn, control the high-power transistor. Now you can see how the biasing resistor keeps the smaller transistor at half power, and how the smaller transistor keeps the high-power transistor in the working circuit at half power, to keep the motor running at half power even when the control generator is not withdrawing electrons.

Again, you may not have been able to understand completely this build-up of circuitry, but don't worry. This is just an illustration of how semiconductor components can be employed in a variety of ways to cope with a variety of conditions. We still have the basic function we started with; all of the additional circuitry makes up an amplifier circuit which will vary the speed of the motor from half power to full power.

Warning: Don't build this or any of the circuits in this book for actual use. They have been simplified for teaching purposes, and lack the refinements necessary to make them perform well.

Naturally, there are many devices we haven't mentioned yet, which can add to the complexity of circuit design. As we come to them later in this book, we will explain how they, too, can be considered in terms of either switching or regulating. For example, we will show how thyristors can be considered simply as combinations of transistors. And we'll show how semiconductor light sensors can be made either to switch or regulate in response to light.

CAN WE SIMPLIFY OUR ANALYSES OF SYSTEMS?

Building-block symbols make an amplifier diagram easier to understand.

There's a simple way of analyzing systems, which we'll call the "building block" method. So many circuits are found within each of the system stages of sense, decide, and act that it's sometimes difficult to conclude anything from a complete circuit diagram that shows how all the components of a system are connected to work together. In practical systems analysis, it's usually enough to think of the circuits in terms of simple building blocks.

In the case of our motor-control circuit, for example, we would think of the entire circuitry within the large triangle simply as an amplifier, without worrying about the components that make it up. Indeed, the building-block *symbol* for an amplifier is a triangle, as we have shown in Fig. 2.8. This circuit is identical to

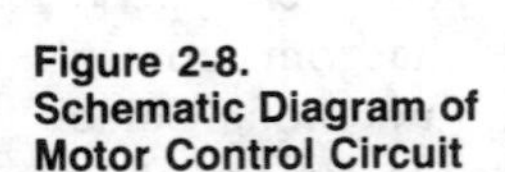

Figure 2-8. Schematic Diagram of Motor Control Circuit

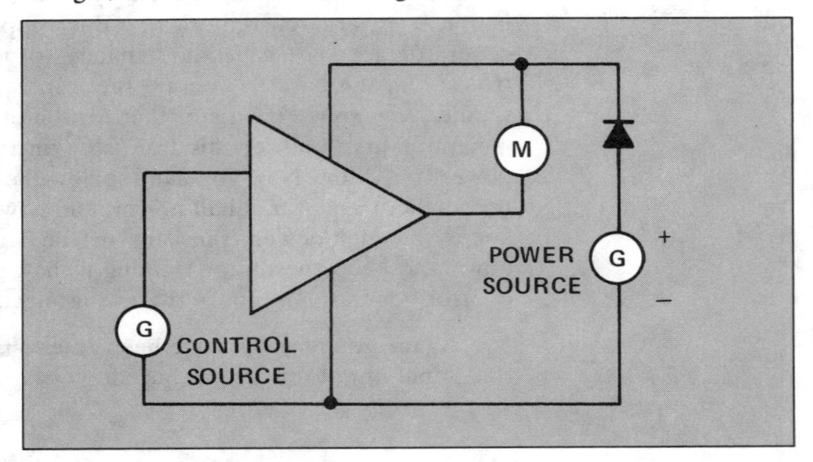

the motor-control circuit shown in Fig. 2.7. This triangle symbol in a system diagram tells us that we have an amplifier, that the wire entering at the left is the control input, that the wire leaving at the right is the output, and that the lines at the top and bottom are the power supply. That's all we really need to know, provided we have the performance specifications for the amplifier. We don't have to be concerned with the *internal* circuitry; in practice, block diagrams don't even show it.

Schematic and system diagrams are simplified in other ways, too. We'll take the circuit of Fig. 2.8 to show how schematics are simplified; although applying these conventions to the simple circuit may appear to complicate it, these same conventions greatly simplify diagrams that may consist of hundreds of lead wires and components. The top and bottom horizontal lines in Fig. 2.8 simply represent the two wires to the power supply, G. In most systems, the power supply produces a constant voltage, say 12 volts. This provides a 12-volt difference between the two connections. This might mean 24 volts on one side and 12 volts on the other, or 18 and 6. But typically, the arrangement is 12 volts on one side and 0 volt on the other.

Earth or chassis ground provides a return path that completes the circuit for electrons to return to the power source.

Zero voltage means electrically neutral, and this neutrality can be obtained by connecting the circuit to the earth — that is, to ground. On the other hand, in trucks and aircraft and other mobile systems where earth ground is not available, the 0-volt side is typically connected to the chassis or frame of the system. Symbols for these ground connections are shown in Fig. 2.9. The

Figure 2-9. Ground Connection Symbols

EARTH GROUND

CHASSIS GROUND

task of the bottom horizontal conductor in Fig. 2.8 is simply to return electrons, at virtually 0 volts, to the power supply. This function can be performed equally well by a wire, the earth, or a chassis.

Similarly, the upper horizontal line is a power-supply line that is also ordinarily represented by symbols. We use a small circle and the symbol "V_{CC}" which stands for power supply voltage.

Vcc is a common identification for the connection to a power source or supply.

Figure 2.10 shows how Fig. 2.8 would look after these conventions have been applied. In many cases, you may not even see the

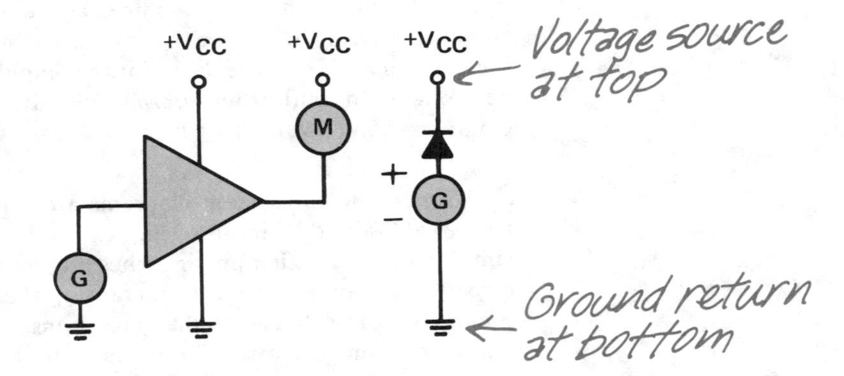

Figure 2-10. Simplified Amplifier Diagram

power-supply and ground connections to building-block symbols, because they are frequently just understood to be there.

This, then, is the typical building-block form of the motor-control circuit we started with. It still tells us everything we need to know for purposes of system analysis: We have a signal generator controlling a motor by means of an amplifier.

WHAT ARE SOME VARIATIONS OF THE BASIC AMPLIFYING FUNCTIONS?

In addition to the triangle symbol for a typical amplifier with one input and one output, you may see variations on this basic symbol. Amplifiers can be designed to amplify in many special ways. For example, the amplifier of Fig. 2.11a is called a "dif-

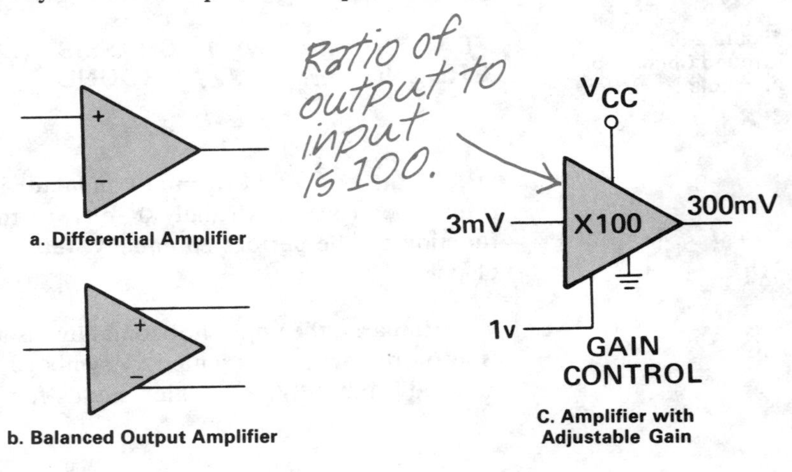

Figure 2-11. Specialized Amplifier Functions

a. Differential Amplifier

b. Balanced Output Amplifier

C. Amplifier with Adjustable Gain

ferential amplifier." It has two inputs, and the output is an amplification of the difference between the voltages of the two inputs. Some amplifiers have differential *outputs,* as shown in Fig. 2.11b; in this case, one output goes high in voltage when the other goes low.

"Gain control" is a means of adjusting an amplifier's output-to-input ratio.

Figure 2.11c shows still another wire to the amplifier, in addition to the power-supply and ground connections. It's labeled "gain control." Since gain control is common to so many amplifiers, we need to discuss it right here. "Gain" is the *ratio* of *output* quantity to *input* quantity of an amplifier. For example, let's say the voltage gain of this amplifier is at 100, as a result of the gain control voltage being at 1 volt. This means that the input voltage will be amplified by a factor of 100. Think back to our loudspeaker system, with the microphone providing the input signal and the output going to the loudspeaker. Suppose a voice signal averages 3 millivolts. Then, the average output voltage would be *300* millivolts, producing a certain loudness of sound. But suppose we want to increase or decrease the loudness. That's where gain control comes in. You can adjust a gain control so that it changes the gain and thereby changes the output. In a loudspeaker system, think of gain control as being adjusted by the volume-control knob. We turn the knob to achieve a signal of 2 volts. The gain goes up to 200. Our output now becomes *200* times 3 millivolts, a much louder sound.

WHAT ARE SOME VARIATIONS OF THE BASIC SWITCHING FUNCTIONS?

There are so many different kinds of switching circuits that there is no schematic outline for them, like the triangle we use for an amplifier. We will just use a rectangular box for our purpose.

Figure 2-12. Basic Switching Circuit

Consider Fig. 2.12. Like an amplifier, a switching circuit typically has a power supply and a ground connection. And it has

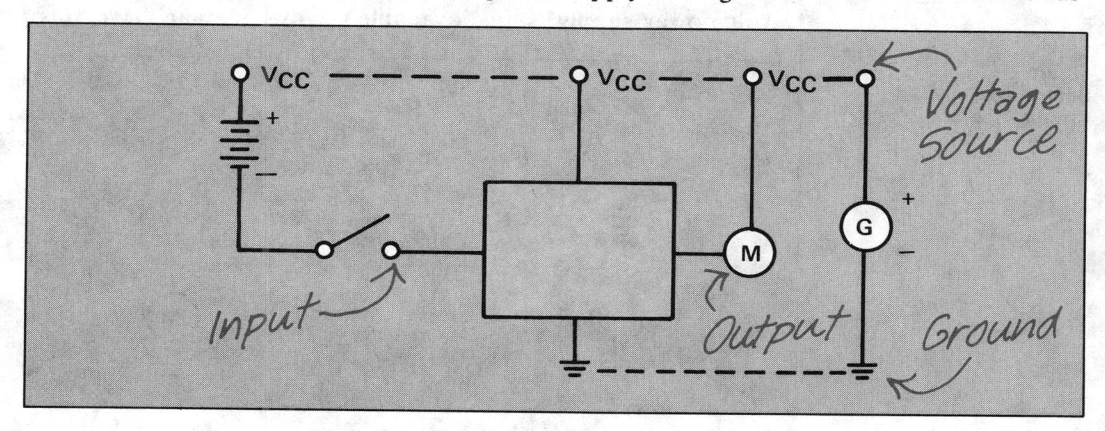

an output connection leading to a load — a motor, "M," in this example — with its other connection going to ground. And we have an input connection at the left, where voltage and current are controlled by some external signal source. In this case, our signal source is a switch and a battery, with the circuit completed through the power-supply voltage. Incidentally, we have followed standard practice in drawing our circuit with input on the left, output on the right, power-supply voltage at the top, and ground connections at the bottom.

Now, what is the function of the building block in Fig. 2.12? Well, the fact is that without being able to see the internal circuit diagram or a list of circuit performance specifications, no one can tell. You would run into the same problem with most building blocks in real systems — except with the simplest amplifiers that are adequately represented by a triangle. A plain box is not sufficient in itself to signify what a circuit does.

To clarify the mystery with respect to Fig. 2.12, let's identify the box as a "current-operated inverting switch." To someone familiar with this terminology, this would mean that when current is withdrawn from the input by the control circuit, current is supplied from the output to power the working device. This function could be performed in a rather crude, limited fashion simply by an NPN transistor, as in the telegraph circuit of Fig. 2.6. But for some other applications of current-operated inverting switches, much more sophisticated circuitry is required.

Switching circuits have numerous applications in digital circuits.

Switching circuits lend themselves to a great many variations, and we will talk about these variations in the next chapter, because switching plays such an important part in digital applications. For the moment, let's examine just one of these variations to suggest how switching circuits can perform additional functions. Suppose we take the basic motor control circuit of Fig. 2.12, and add a *second* power-supply voltage, of 6 volts, as shown in Fig. 2.13. The 12-volt power supply is still available. Now, we can have this

Figure 2-13. Simplified Switching Circuit Diagram

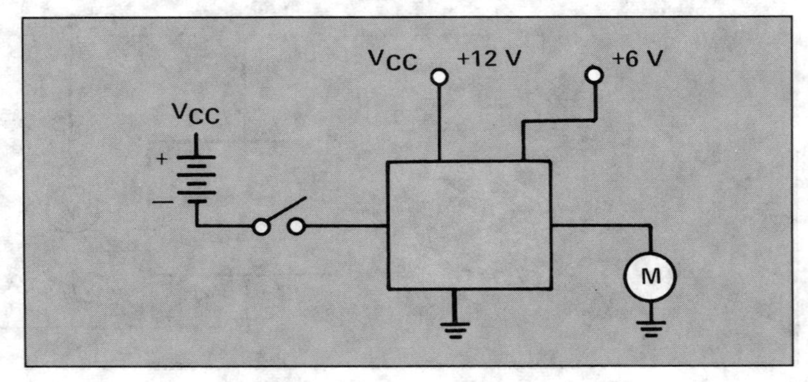

circuit designed so that we can switch *between* the 6-volt and 12-volt supplies as desired, changing the speed of the motor between low and high levels.

WHAT ABOUT OSCILLATORS, MODULATORS, AND OTHER BUILDING BLOCKS?

Switching and amplifying circuits are fundamental to all types of electronic circuits.

We've talked about amplifier circuits and switching circuits as the building blocks used to build stages of a system. But you've probably heard of other types of building blocks — oscillators, mixers, rectifiers, modulators, detectors, and others. You may have the feeling that we're not telling you the whole story, and that electronic circuits cannot be as simple as just amplifying and switching. Yet in truth, they are. Even though we have a great many functions that bear different names, they can still be considered simply as *variations* of the basic functions of switching and amplifying.

An oscillator is a type of amplifier that uses feedback to sustain oscillations.

Take an oscillator, for example. Figure 2.14 is an oscillator—drawn in building-block style. The oscillator circuit, represented by the box, has a power-supply voltage connection at the top, a ground connection at the bottom, and an output at the right. The current and the voltage in the output line are fluctuating in a regular, predictable manner. The output of an oscillator is typically a sawtoothed-wave pattern or a smooth sine-wave pattern; we've shown both of them.

How are such regular fluctuations produced? The oscillator is made of an amplifier connected to power supply and ground; but the output is fed back to what we call a "non-inverting input."

Figure 2-14. Oscillator Circuit

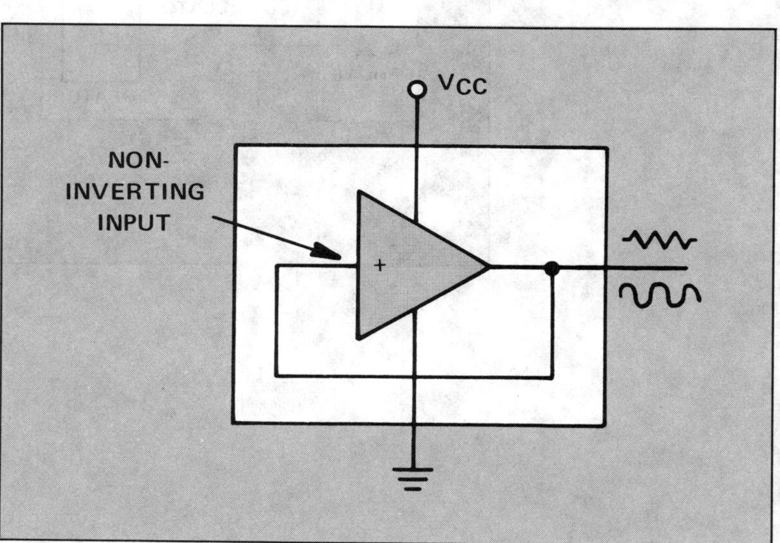

The amplifier is designed so that an increase in the input voltage causes an increase in output voltage. Similarly, when the input voltage decreases, so does the output voltage. Suppose we begin with the output voltage increasing. This output is fed back to the non-inverting input, causing the input to increase. This process continues until the circuit limits the output to some voltage. At this point, the process reverses, with the output and input decreasing until a lower limit point is reached. The result is a series of oscillations in output voltage and current. The shape of the oscillations depends on details in the design of the amplifier. In particular, the shape depends on the time delay between an input and the response at the output of the oscillator.

A modulator is a type of amplifier that impresses information (such as speech or music) on high-frequency waves produced by an oscillator.

Let's look at still another example of a circuit function that is a variation on the amplifier, the modulator. You may be aware that the modulator is the important circuit in an AM radio transmitter. To explain the modulator function, we have built up a diagram of an entire radio transmitter system, in Fig. 2.15. The system includes an oscillator, which we have just discussed. Now, let's see how the building blocks are put together.

Figure 2-15. Radio Transmitter Diagram

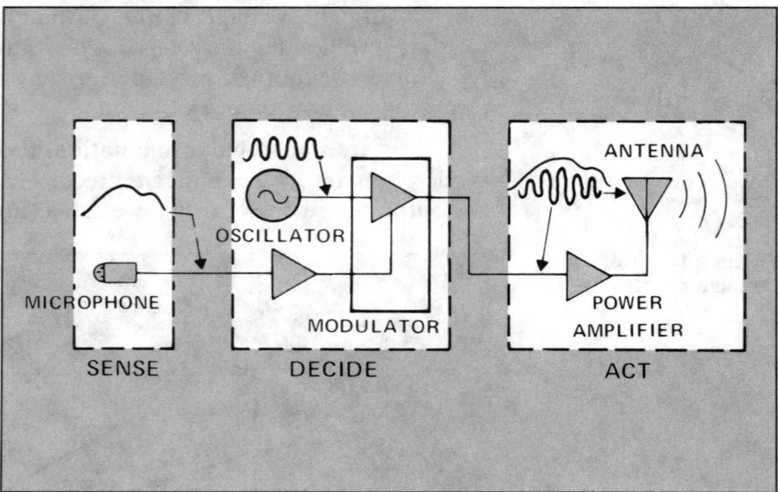

Suppose we are broadcasting the sound of a voice received by a microphone. The microphone produces fluctuating voltages and currents which duplicate the action of the sound waves. We have shown the wave configuration for a very brief fragment of speech, as it leaves the microphone lead. These electrical waves have frequencies of about one kilohertz — a thousand cycles per second. But the production of radio waves by a broadcasting antenna requires that we produce alternating current in the antenna at frequencies of about 1 megahertz — a *million* cycles per second. So it's obvious we cannot simply broadcast the waves produced by the microphone.

The oscillator we've just talked about, however, *can* produce high-frequency waves. As we saw, it produces these waves in a regular repetitive fashion. All of the oscillator's waves have the same amplitude, or height. So we have added an oscillator to our radio transmitter. And to keep its function in mind, we have shown the wave configuration it produces, a series of regular waves of the same amplitude.

Obviously, these repetitive waves contain none of the information we wish to transmit. And the waves from the microphone, full of information, can't be broadcast. This is where the modulator comes in. We have drawn it as a simple box, with power-supply and ground connections understood. Regardless of what circuitry is inside the box — whether it's an integrated circuit, or discrete transistors and other components wired together, at its heart there is an amplifier function. So we've shown the amplifier portion of the modulator, to make clear what happens.

In this example of amplitude modulation, the modulator controls the gain of the amplifier.

The oscillator is connected to the input of the modulator's amplifier. The microphone's output is connected to control the gain of the amplifier. Now, the voltage produced by the microphone will regulate the high-frequency waves from the oscillator. When the voltage wave produced by the microphone is high, the waves produced by the oscillator will be amplified proportionally. When the microphone voltage is lower, the waves from the oscillator will be lower. In this way, the waves from the microphone modulate the waves from the oscillator. The result at the output is waves that look something like the wave shape shown — they have the high frequency of the oscillator, and the amplitude of the signals from the microphone. *They can* be broadcast.

To complete the diagram, we have supplied a power amplifier to give the signal the great strength required for broadcast, and a broadcast antenna.

The radio transmitter is readily analyzed into the system stages of sense, decide, and act, as shown in the figure. The *sense* stage is the microphone, which senses the incoming sound waves and converts this information into electrical form. At the other end of the system, the power amplifier and antenna make up the *act* stage, converting low-power electrical information into properly modulated high-frequency radio waves for reception at distant points. The other building blocks in between form the *decide* stage, which manipulates the information as required.

This anatomy of a radio transmitter system provides an example of how system stages are made up of building blocks called "circuit functions." It also demonstrates that, even though circuit functions have names such as "oscillator" and "modulator," these are really just variations of the basic amplifier. Once again, we have shown that all circuit functions can be classified either as amplifying types or switching types.

There were no *switching* circuits in this radio transmitter example. In the next chapter, we turn our attention to the details of switching circuits, with emphasis on digital logic circuits, the kind that make the modern computer possible.

Quiz for Chapter 2

1. The operation of a transistor is an example of the basic function of all semiconductor devices, which is:

- ☐ a. Switching and regulating the flow of electrons
- ☐ b. Too complex to understand
- ☐ c. Raising the voltage of the circuit
- ☐ d. Withdrawing electrons
- ☐ e. c and d above

2. The three basic regions of a transistor are:

- ☐ a. Collector, bias, emitter
- ☐ b. Emitter, base, conductor
- ☐ c. Emitter, base, collector
- ☐ d. Cathode, grid, plate
- ☐ e. None of the above

3. The schematic symbol of an NPN transistor is:

- ☐ a. 
- ☐ b. 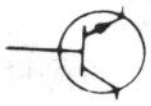
- ☐ c. 
- ☐ d.

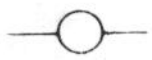

4. When a transistor regulates current, it is said to be "amplifying" current because:

- ☐ a. The collector "collects" electrons
- ☐ b. The voltage at the emitter is greater than the voltage at the base.
- ☐ c. The current in the working or output circuit is greater than the current in the control or input circuit
- ☐ d. None of the above
- ☐ e. All of the above

5. A transistor can act as a switch if it is especially built and connected so that it:

☐ a. Provides a voltage increase at the output which is directly proportional to the signal at the input
☐ b. Allows the current crossing its emitter to be lower than its output
☐ c. Either allows full power to flow in the working circuit, or allows no current at all to flow in the working circuit
☐ d. None of the above
☐ e. All of the above

6. Electrons flow through the working or output circuit from a power source and typically return to the power source via:

☐ a. The generator
☐ b. The base connection of a transistor
☐ c. A wire assumed to be constantly at zero voltage, called "ground," which often is actually the earth
☐ d. A rectifier diode which converts ac to dc
☐ e. None of the above

7. Regardless of the name given to an electrical circuit, it can be classified as:

☐ a. Either a switching circuit function or an amplifying circuit function
☐ b. Either an input circuit or an output circuit
☐ c. A modulator
☐ d. An oscillator
☐ e. a and b above

8. "Building blocks" in an electric system refer to:

☐ a. A way of assembling circuits in packages that are plugged together like a child stacks toy blocks
☐ b. Portions of the system that for purposes of systems analysis can be thought of as boxes with inputs, outputs, and power supply connections, without studying circuit details inside
☐ c. Circuits that are as "simple as toy blocks"
☐ d. None of the above

9. The "building block" symbol for an amplifier is:

- ☐ a. 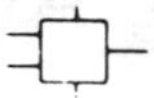
- ☐ b. 
- ☐ c. 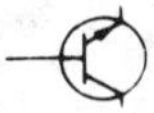
- ☐ d. 
- ☐ e. None of the above

10. When you encounter an unfamiliar circuit function (that is, a building block in a system), the first information you should ascertain in understanding it is:

- ☐ a. How many components does the circuit have?
- ☐ b. Does it use ac or dc?
- ☐ c. What does the circuit cost?
- ☐ d. Is it a switching or an amplifying type circuit?
- ☐ e. b and c above

11. The specification called "gain" is defined as:

- ☐ a. The difference between input voltage and output voltage of a transistor
- ☐ b. The ratio of output quantity to input quantity of an amplifier
- ☐ c. The total increase in output quantity over the input quantity of an amplifier
- ☐ d. Input quantity to an amplifier divided by output quantity
- ☐ e. None of the above

12. The electron flow in the working circuit passes through an NPN transistor from:

- ☐ a. Base to emitter
- ☐ b. Collector to base
- ☐ c. Collector to emitter
- ☐ d. Emitter to collector
- ☐ e. Emitter to base

Key Words

AND Gate
Gate
Half-Adder
Logic Gates
NAND Gate
NOR Gate
NOT Gate
OR Gate

Definitions are found in the glossary in the back of the book.

How Circuits Make Decisions

Most of this chapter is devoted to the *decide* section of our Universal System. In the last chapter, we discussed switching and amplifying circuits in rather general terms, and we didn't specify exactly how these circuits fit into the three stages of sense, decide, and act. The fact is that *circuit functions* do not sense and they do not act. The sensing and acting are done by *devices.* The microphone, for example, is a sensing *device.* The motor is an acting *device.* In any system, you might very well find switching and amplifying circuits in the sense and act stages, but they do not perform the *function* in either kind of stage. In the sense stage, they merely translate the output of the sensing device into some form of information. In the act stage, they provide power for the acting device in response to information from the decide stage. So we can say that *circuits* do not sense and they do not act.

But switching and amplifying circuits *can* make *decisions.* Every day in this country, in factories and offices and computer centers, literally trillions of decisions are being made by electrical circuits, every second. These decision-making circuit functions are the brains of the *decide* process in virtually every electronic system. This chapter is concerned with how circuit functions make those decisions.

We have learned a little about decisions in an electric system already. We know that they are made by the decide stage. They require information inputs, and produce new information outputs. In Chapter 1, we said there were two ways to get this electrical information in and out — by the analog method or the digital method. So we can conclude that there are two kinds of decisions — analog decisions and digital decisions.

Most of this chapter deals with digital decisions, because they are somewhat more complex than analog, and because they are at the heart of so many pieces of electronic equipment, from simple toys to super computers. But before tackling digital decisions, we should talk about analog decisions briefly.

HOW DO AMPLIFYING (REGULATING) CIRCUITS MAKE DECISIONS?

Analog systems can make decisions based on different values of voltage, current, or frequency at their inputs.

Recall that in Chapter 1 we spoke of voltage analog, current analog, and frequency analog. We saw that code systems can be devised that employ differences in voltage, current, and frequency. Take voltage analog, for example. We transmit varying voltages, and they are received and recorded at a reception point. Suppose the voltage levels for a transmission looked like Fig. 3.1. If we measured the heights on the voltage

Figure 3-1. Analog Voltage Waveform with Different Levels

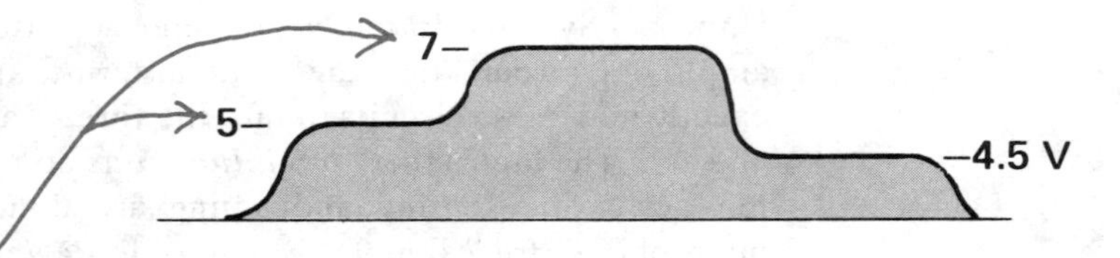

curve, we would find that we have 5 volts, followed by 7 volts, followed by 4.5 volts. If we had arranged the code to let the voltages stand *directly* for numbers, our transmission here would say "5, 7, 4.5." On the other hand, if we had arranged that our code would mean the *square root* of the numbers to be communicated, the message would read "25, 49, 20.25."

Current analog works much the same way, the difference being that current analog depends on changes in current instead of changes of voltage. By the same token, *frequency* analog transmits information based on changes in the frequency of alternating current.

Analog methods are used for many purposes other than the transmission of numbers, of course. The loudspeaker system we saw earlier is an analog system in which varying voltages represent fluctuating air pressure. Telephones carry our voices by analog techniques, and radio transmitters are also analog systems.

It's readily obvious that there is a transmission of information in analog systems, but are *decisions* really being made? Yes. In fact, we have already seen how this is done. Think back to the radio transmitter. We said that the decide stage was the modulator, which is an amplifier. The input is high-frequency voltage waves from an oscillator. A voltage signal from the microphone controls the gain of the amplifier. The amplifier multiplies the input voltage by the gain to produce the output voltage. Suppose that at a given instant, we have a gain-control voltage of 2 millivolts producing a gain of 2. At the same instant, suppose the input voltage is 3 millivolts. The output, then, would be 3 multiplied by 2 to produce 6 millivolts. Figure 3.2 makes this

Figure 3-2. Amplifier with Gain Control

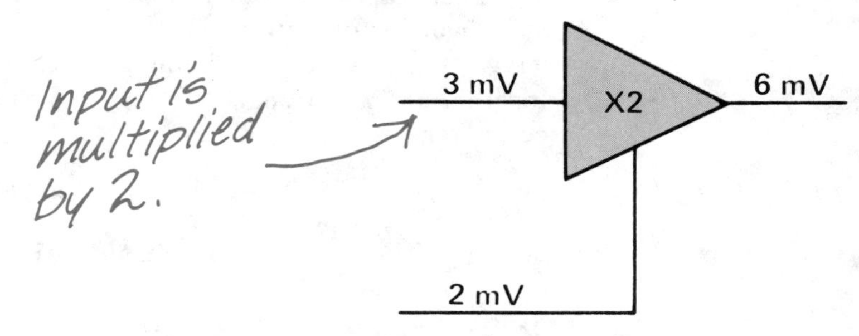

clear schematically. This combining of input information to produce new information at the output is clearly a decision. You can see that amplifiers continuously make decisions.

Analog circuits can perform mathematical operations such as multiplication, addition, subtraction, division, integration, and differentiation.

Take a differential amplifier. We said that the output voltage of a differential amplifier is an amplification of the *difference* between the input voltages. Consider Fig. 3.3 — this shows a

Figure 3-3. Differential Amplifier with Gain Control

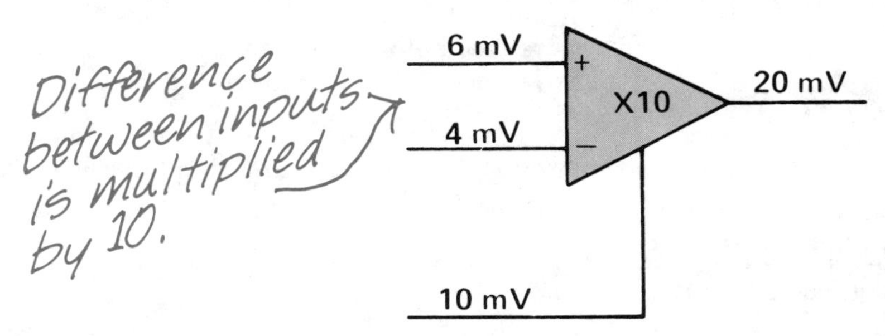

differential amplifier experiencing inputs of 6 millivolts and 4 millivolts; it has a fixed gain of 10. Six minus 4 equals 2; times 10, it equals 20, yielding an output of 20 millivolts. Here again, we have an example of decision making — in this case, a decision that incorporates both subtraction and multiplication.

Figure 3.4 shows addition by means of current analog. Suppose one input is 6 milliamps and another input is 2 milliamps.

Figure 3-4. Analog Addition Using Currents

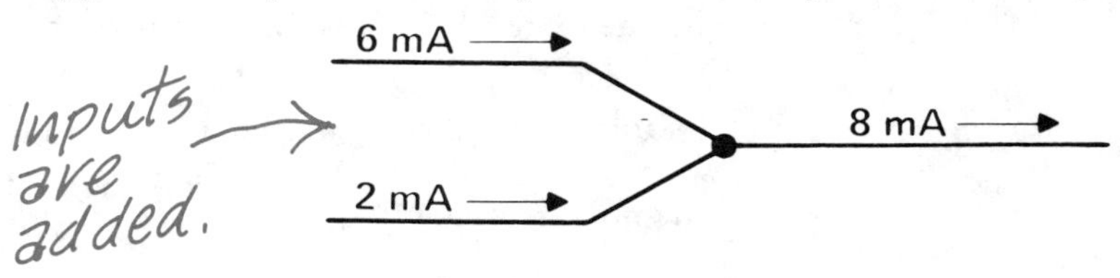

To add these two, we simply run the two currents together and measure the resulting current with an ammeter. In this case, it would read 8 milliamps.

There are still other analog circuits that perform other mathematical operations including division, finding logarithms, integrating, and differentiating. When we string these circuits together in the proper combinations, the result is an analog computer.

Now that you are familiar with the electrical basis for analog decisions, the rest is simply a matter of mathematics and circuit design. The rest of this chapter will be devoted to *digital* decisions.

HOW DO SWITCHING CIRCUITS MAKE DECISIONS?

Digital systems make decisions by using binary switching circuits that are always one of two values, either on or off.

Let's consider the digital computer for a moment, since it's the most sophisticated type of digital system. What is it in a computer that makes decisions? It's simply thousands upon thousands of switching circuits — simple little circuits that can do no more than turn on and turn off. Yet a computer can perform tremendously complex mathematical operations at blinding speed. It should be obvious that, somehow, computer science has found ways to employ this simple on-off function of switches to solve extremely difficult problems.

Think in terms of information. How much can we say with a switch? Figure 3.5 is the simplest sort of circuit, with a power

Figure 3-5. The Basic Activity of a Digital System: Switching On and Off

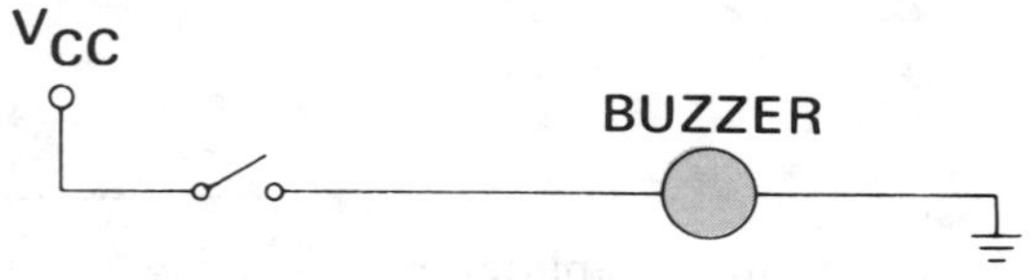

V_{CC}

BUZZER

source, a switch, and a buzzer, completed to ground. What information can we send from the switch to the buzzer? We can close the switch and allow the high voltage of V_{CC} from the power source to pass through the wire and sound the buzzer. Or we can open the switch and allow the voltage in the wire to return to ground, so the buzzer no longer operates. That's all we can say with the switch. We are limited to two statements — no more than that.

If your vocabulary were to be limited to only two words, what two words would you choose? The wise man would choose "yes" and "no." You would get along better with those two words than with any other two that you could choose. In digital electrical systems, we assign the same meanings to the two available voltage levels, and we normally let high voltage stand for "yes" and low voltage stand for "no."

Just by using multiple binary switching circuits in unique combinations, digital systems can process extremely detailed information.

Digital information is based on this simple yes-no principle. But it's not as restrictive as it might seem. In fact, it is infinitely versatile. Let's go through a little exercise that illustrates this. Your friend is thinking of one card in a deck of playing cards. You are to find out what that card is. But the other person is limited in his response to only the words "yes" and "no." How do you find out which card your friend has selected? You could start by simply guessing at random. With random guesses, you might be right on the first try — or it might take you 51 tries. But by using a system, you can guarantee that you will select the right card by asking no more than 6 questions. Let's try it: Your friend is thinking of the Jack of Diamonds. You ask, is the card black? Answer, no. Now you know the card is red. You ask, is it a heart? Answer, no. Now you know the card is a diamond. You ask, is it below 8? Answer, no. Now you know the card must be an 8 or above. You ask, is it below Jack? Answer, no. So the card must be Jack or above. You ask, is it below King? Answer, yes. Therefore, the card is either Queen or Jack. You ask, is it a Queen? Answer, no. The card must be the Jack of Diamonds.

So you can see that detailed information *can* be communicated using only the words "yes" and "no" — or said electrically, with high voltage or low voltage. So much for *information.* Now, what about *decisions?* Let's look at an example of a digital decision.

Figure 3-6.
Digital Decision-Making in a Thermostatic Control System

Imagine the thermostatic control system for a central heating unit, one that burns gas. Consider Fig. 3.6. The thermostat on the

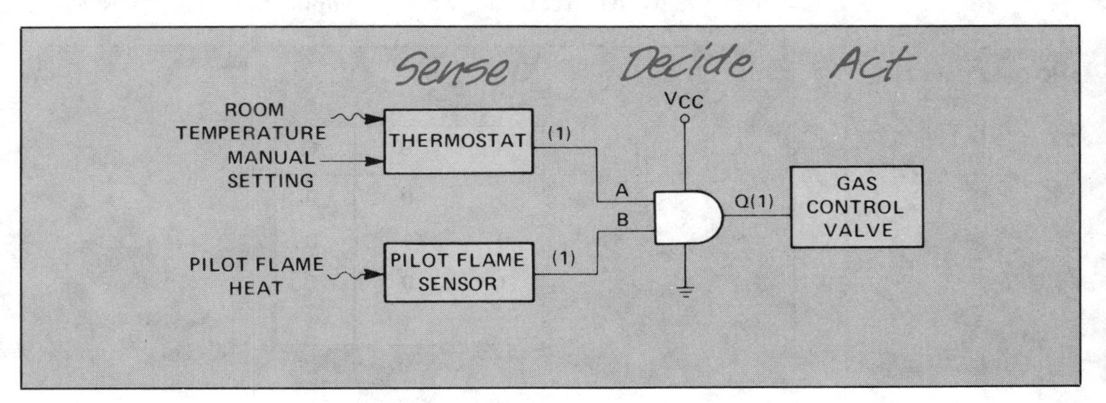

wall compares actual room temperature with the desired temperature setting you have dialed in. The output wire from the thermostat carries digital information. A high voltage means "yes, the room needs more heat." A low voltage means "no, the room does not need heat." This signal is sufficient to turn the gas valve on or off at the proper time.

However, as a safety factor, the system must incorporate a *second* stream of information. We need a temperature sensor next to the pilot flame which will determine whether this flame is on or not, because we depend upon it to ignite the main burner. In an all-electronic system, the information from this sensing function would be either a high voltage saying "yes, the pilot flame is burning," or a low voltage saying "no, it is not burning."

The switching circuits that make decisions in digital systems are called logic gates.

Based on these two inputs of information, a decision must be made. To accomplish this, we have designed into our all-electronic system a decision-making switching circuit called a "logic gate." There are three basic types of logic gates; the one we are using is an AND gate. Its electronic symbol is drawn as shown in Fig. 3.6. The AND gate has an output which goes to the gas valve control. If the thermostat says "yes, we need heat," AND the pilot sensor says "yes, the pilot is burning," *then* the AND gate decides "yes, turn on the gas valve." On the other hand, if we get a "no" at either of these inputs, then the output will be "no." Using an AND gate, we get a "yes" output *only* if *both* inputs are "yes."

Figure 3.7 is what logicians call a "truth table," which shows clearly and precisely all of the possibilities for an AND gate. We have named our two inputs A and B, and the output Q. The truth table has three columns: One column for each input and one for the output. Instead of writing "yes" and "no," the customary symbols in digital logic are 1 and 0; 1 stands for yes, 0 stands for no. Remember we're using them only as *symbols,* not as numbers.

Now, we can read the table. If input A is one ("yes") and

Figure 3-7. AND Gate Truth Table

The AND Gate's output goes high when Inputs A and B are high.

AND Gate Truth Table

A	B	Q
0	0	0
0	1	0
1	0	0
1	1	1

input B is also one ("yes"), we get a Q "output" of one ("yes"). Let's look at other possible combinations: If A is one and B is zero, we get a zero output, because the AND gate will give a "yes" answer only when we have "yeses" at *both* inputs. If A is zero and B is one, the same is true; the output is zero. And obviously, if A and B are both zero, then the output is zero.

The truth table presents all possible combinations and the decisions that result. You can check it against your knowledge of the real world. The bottom row, for example, indicates that the thermostate says "yes, we need heat" and the sensor says "yes, the pilot is on" so the decision is "yes, turn on the gas." The case shown in the third row is that input B says "no, the pilot is off," so the decision is "no heat." In the second row, input A says "no, it's plenty warm enough," so the decision is "no heat." The case of the first row is even more obvious: The house is warm enough, and the pilot is out anyhow, so the decision is "no heat."

HOW DO LOGIC GATES WORK?

Mechanical switches and relay contacts that form AND and OR logic circuits can be used to illustrate how logic gates work.

Most logic gates are semiconductor circuits, but gates can be built using electro-mechanical relays, and since they are a little easier to understand, we will use them to illustrate the principle of the AND gate. A relay is a simple mechanical switch that remains in the off position until a voltage is applied to its electromagnet, turning it on.

The circuit in Fig. 3.8 uses relays. One of the relays is powered by input A, and the other by input B. The main circuit has a power supply, and its output is Q. You can see how the AND gate works. Close only switch A, and electricity will not flow

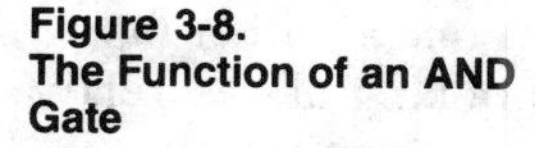
Figure 3-8. The Function of an AND Gate

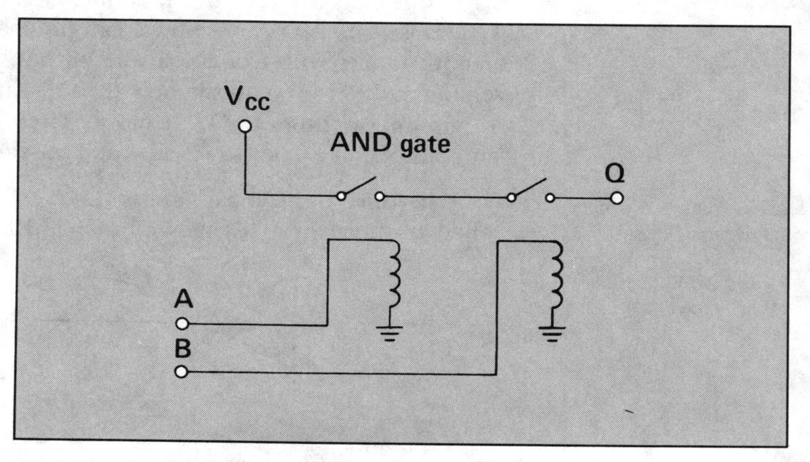

because the other switch is open. Close B and leave A open, and still no flow. But close them both and you'll get a "yes" signal at the output. This is a simple example of how an electrical switching circuit can make a digital decision.

The three basic types of logic gates are the AND, OR, and NOT gates.

There are three basic logic gates: The AND gate, which we have just seen, the OR gate, and the NOT gate. These three gates are sufficient to solve any mathematical problem, if enough of them are put together in the right combination. A little later in this chapter, we will show how digital computers perform mathematical operations using these gates.

But first, let's get acquainted with the other logic gates. An OR gate is shown in Fig. 3.9, along with its truth table and a relay

Figure 3-9. The Function of an OR Gate

Truth Table

A	B	Q
0	0	0
0	1	1
1	0	1
1	1	1

The OR Gate's output goes high when Input A, Input B, or both go high.

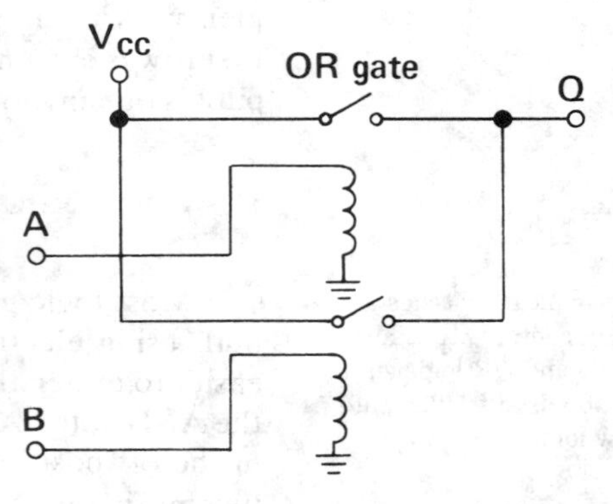

version. We can check its function by reading the truth table. An OR gate is so-called because it will produce a "yes" output when either one OR the other of its inputs is "yes." Thus, in the truth table, if A is one and B is one, Q is one. If A is one but B is zero, Q is still one. If A is zero but B is one, Q is one. *Only* if A is zero AND B is zero, is Q zero. So we can say that a one at either input or both inputs will produce a one output. Looking at the relay version makes it easy to see how the OR gate works. Close switch A, and energy flows to Q. Or close switch B, and the same thing happens. Or close both switches, and the same thing happens.

A NOT gate has only a single input.

The third basic logic gate is the NOT gate, which is usually called an "inverter." It's shown in Fig. 3.10. Note the circle at the

Figure 3-10. Inverter Symbol

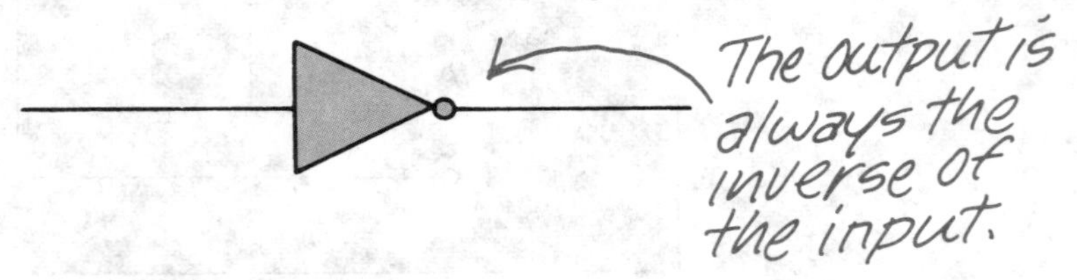

apex of the triangle, indicating negation. Note also that the NOT gate has only *one input.* The function of the NOT gate is simple: Its output is always the *opposite* of its input. A "yes" at the input results in a "no" at the output; a "no" at the input results in a "yes" at the output.

All logic circuits are simply combinations of the three basic functions AND, OR, and NOT. Two of the possible *combinations* are so widely used that we should be familiar with them, too. The NAND gate symbol is shown in Fig. 3.11. It is made up of an AND

Figure 3-11.
NAND Gate Symbol

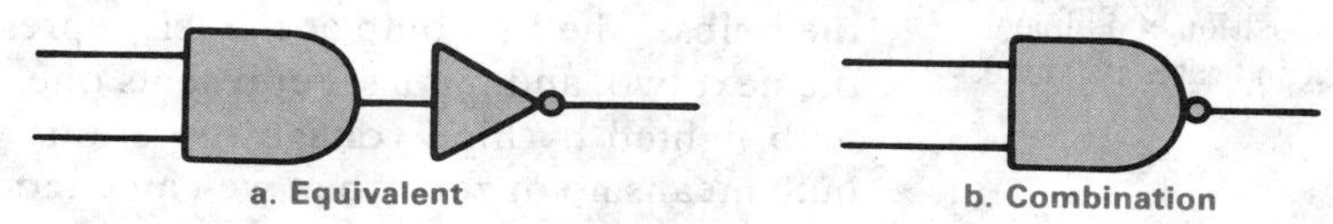

gate followed by an inverter, so we use the symbol for an AND gate, plus the circle meaning NOT. "NAND" means NOT-AND. With the NAND gate, if you have "ones" at both inputs, the "one" *output* is inverted to a zero.

The NOR gate is similar. Its symbol, shown in Fig. 3.12, is simply the OR with a circle at the tip. NOR means NOT-OR. It is

Figure 3-12.
NOR Gate Symbol

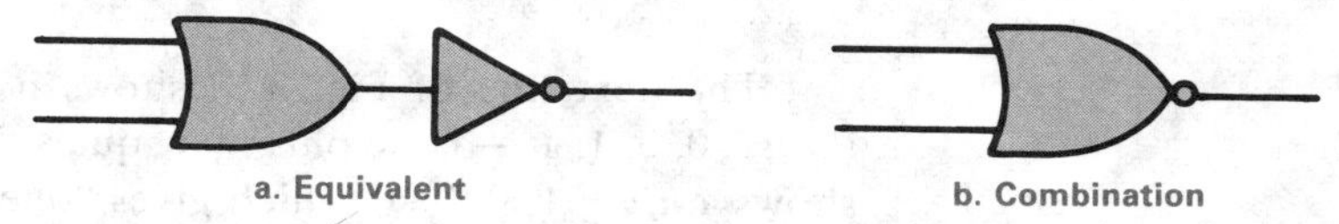

made up of an OR gate followed by an inverter. So if we have a zero and a one at the inputs, the "one" at the *output* is inverted to a zero.

Any type of logic gate except the NOT gate can have several inputs.

Except for the NOT gate, any of these gates can have more than two inputs. If we had, for example, three inputs to an AND gate, we would have to have ones at all three inputs to get a one at the output. Three inputs to an OR gate, similarly, would mean that the output is one if any of the three inputs is one.

HOW IS NUMERICAL INFORMATION TRANSMITTED IN ELECTRICAL CIRCUITS?

Now that we know how basic logic gates work, we're ready for our next step toward putting them together to solve complex mathematical problems. The next step is to see how we are able to transmit numerical information using only one and zero. We mentioned the binary number code briefly in an earlier chapter. Since we will be using this code extensively, let's review it.

Let the circles in Fig. 3.13 represent four light bulbs that can be turned on and off individually by supplying high voltage or low

Figure 3-13. Binary Numbers

⑧④②①

0 1 0 1 = 5
0 0 1 1 = 3
1 0 0 1 = 9

As in the decimal system, the position of a binary digit indicates its value.

voltage to them. We assign a different numerical value to each of the bulbs. The first bulb at the left represents eight, the next four, the next two, and the last represents one. If we agree that an "on" bulb lighted by high voltage represents yes or one, and an "off" bulb means no or zero, then we can encode any number from zero to fifteen.

For example, if we turn on the second and fourth bulbs, our message would read: 0, 1, 0, 1. Reading from right to left, the message says: "yes, we have a one; no, we don't have a two; yes, we have a four; no, we don't have an eight." Summarizing the message, we have "one plus four" — so the number transmitted is "five."

The next line of Fig. 3.13 shows how we would transmit a *three:* 0, 0, 1, 1 — "one plus two equals three." And the last line shows a *nine;* 1, 0, 0, 1, which gives "one plus eight equals nine."

HOW DO WE ADD BINARY NUMBERS?

Now that we can convert any number to binary code, let's see how we can add binary numbers. The rules for adding binary digits are just a little different from the decimal system. The only three possibilities are shown in Fig. 3.14. Zero plus zero equals

Figure 3-14. Binary Addition

0	1	1
+0	+0	+1
0	1	10

zero. One plus zero equals one. But one plus one, though it equals two, yields the *binary form* of two, written "one-zero;" the one is carried to the next column. As you can see, these calculations are simple enough to be done by electronic gates.

Binary numbers are added by summing the values of the digits in corresponding digit positions, just as with decimal numbers. Carries from one digit position to another are included as they are generated.

Now we can set up an addition problem as in Fig. 3.15. We have shown two inputs; A equals 12, B equals 14. We will add these together to get the sum, 26. We have placed the binary digits so that they fall in the columns that show how much each is "worth." Now we're ready to add, beginning with the column at the right. In the first column at the right, zero plus zero equals zero. In the second column, zero plus one equals one. In the third, one plus one equals two, but we write it in binary numbers as one-zero; so we write zero in the third column and *carry* the one to the top of the fourth column.

In the fourth column, we now have three digits to add, and we do this in two steps. In the first step, one plus one is one-zero, and this one is carried to the "sixteen" column. In step two, the zero we have just produced added to the one below it, gives one. Finally, the one in the "sixteen" column is brought down as part of the sum.

Now we have the answer: 1, 1, 0, 1, 0. We can interpret this as "yes we have a sixteen, yes we have an eight, no we don't have a four, yes we do have a two, and no we don't have a one." The result is 16 + 8 + 2 equals 26.

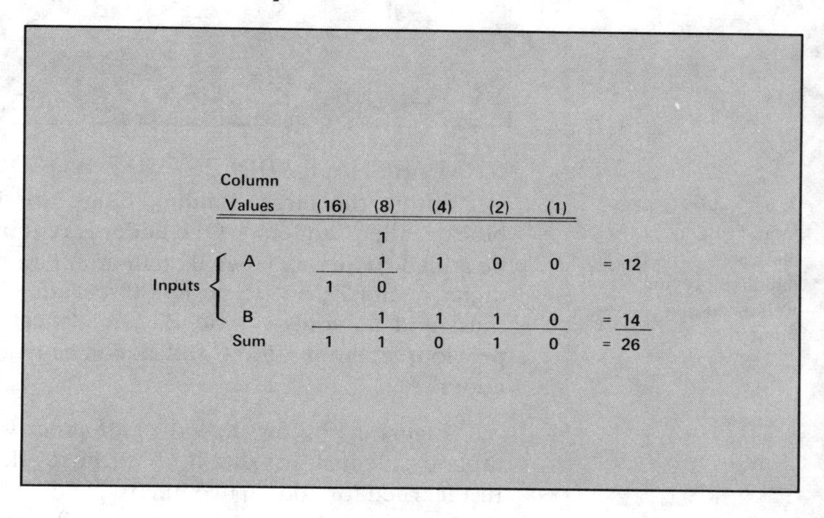

Figure 3-15. Binary Addition with Carries

Now, let's see how a computer does that same problem using gates. Considering the entire circuit of Fig. 3.16 as a building block, we have four wires inputting our first number, number A, which is 1, 1, 0, 0, or 12. And we have four wires inputting our number B, which is 1, 1, 1, 0, or 14. Finally, we need five output wires for our sum, which is 1, 1, 0, 1, 0, or 26.

Figure 3-16. Binary Addition Using Logic Gates in Adders

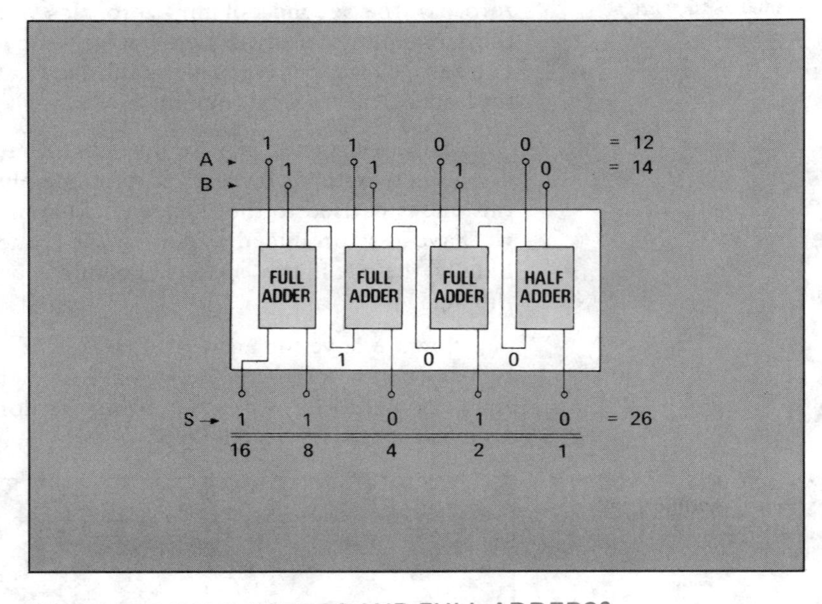

WHAT ARE HALF-ADDERS AND FULL-ADDERS?

A half-adder can add two inputs to generate a sum and carry. A full adder can handle a previous carry as well as adding the two inputs.

Within the large building block are four smaller building blocks called "adders." One adder is required for each column to be added. The one labeled "half-adder" merely accepts the input digits A and B; but the other three adders, called "full-adders," must add not only A and B, but also a digit carried from the previous column. Each full adder can be made from two half adders.

Figure 3.17 shows how the half-adder is constructed; the truth table beside it shows that it conforms to the simple rules we have just learned for adding two binary digits. It's worthwhile to review the four cases: When A is 0 and B is 0, we want a sum of 0 and a carry of 0. When either A or B is 1, we want a sum of 1 and a carry of 0. And when both A and B are 1, we need a sum of 0 and a carry of 1.

Adder circuits contain both AND and OR gates, as well as inverters, to provide the correct outputs for all possible combinations of inputs.

Now, consider how the gates of Fig. 3.17 are interconnected to produce these outputs. We have two AND gates, an OR gate, and an inverter. We'll test the circuit to make sure it works right. Suppose we take the first line of the truth table, and add 0 plus 0. When both inputs to an OR gate are 0, you'll recall, the output is 0. This is the 0 that becomes one of the inputs to the lower AND gate. At the same time, both 0's are also fed into the AND gate. Since we do not have a 1 and a 1, the output is 0 — so our carry is 0. But that 0 is also fed through the inverter, turning it into a 1, and inputting it to the lower AND gate. Thus, our inputs to the lower AND gate are 1 and 0, so the sum is 0. The result is that our input of 0 and 0 results in a carry of 0 and a sum of 0, exactly as the truth table says it must.

Figure 3-17. The Construction of a Half-Adder

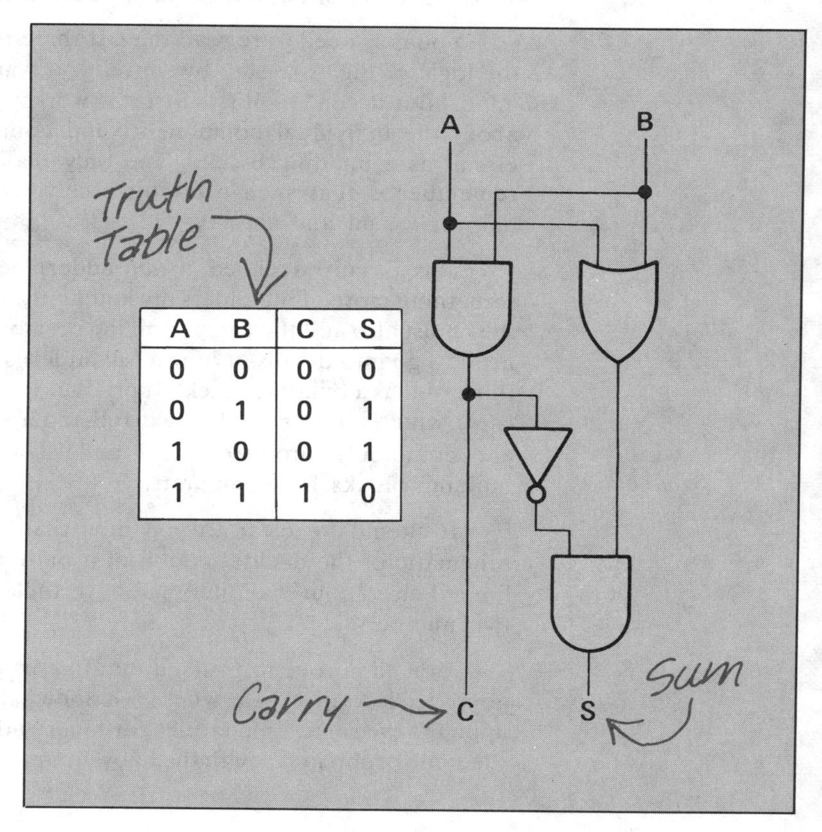

A	B	C	S
0	0	0	0
0	1	0	1
1	0	0	1
1	1	1	0

Let's try another combination, a 1 and 0. Since one of the inputs to the OR gate is a 1, the output is 1. But since the inputs to the AND gate are not *both* 1 and 1, *its* output is 0 — so the carry is 0. This 0 also passes through the inverter, and becomes a 1 input to the lower AND gate. So the inputs to the lower AND gate are 1 and 1, giving us a sum of 1. Here again, the carry of 0 and sum of 1 match our truth table.

The final case takes inputs of 1 and 1. Since at least one of the inputs to the OR gate is 1, its output is 1. Since both inputs to the AND gate are 1, its output is also 1. So the carry is 1. When this 1 is inverted, it inputs a 0 to the lower AND gate, and since the inputs to the lower AND gate are not *both* 1 and 1, the sum is 0. Thus, the carry of 1 and the sum of 0 satisfy the truth table.

You may need to re-read the last three paragraphs and follow the logic of Fig. 3.17 step by step if you want to get it clearly in mind. But if you're satisfied that it works, you can really forget about the individual components and continue to think of the circuit as a building block. The only really important thing to remember is that such a circuit receives *inputs* A and B, and *outputs* a sum and a carry.

This circuit is called a *half*-adder, because *full*-adders do something more. Full-adders must not only accept inputs A and B; they must also add the carry from the previous column. There's no need to go into details of how a full-adder is interconnected. Just think of it as a building block. It produces a sum, and it produces a carry which goes on to the next full-adder. Half- and full-adders are commonly produced as standard integrated circuits — building blocks for computers.

It should be clear to you now that Fig. 3.17 is really the schematic of the decide section of a baby computer. It's a very limited circuit but it's functional nevertheless, because it will add two numbers.

You can prove to yourself that this arrangement will work by going back to Fig. 3.15, where we added the numbers manually; applying the same simple rules for binary addition, you can follow the same problem through the baby computer of Fig. 3.16.

HOW CAN WE SPEED UP ADDITION?

The adder of Fig. 3.16 performs 4-bit binary addition just as we do. It gets the carry for each digit position from the summing of the two digits to the right. This means that the circuit cannot decide on the sum and carry out of the second digits until the rightmost digit sum has been completed. Similarly, the summing of the third digits must wait for the carry to be generated by the summing of the second digits from the right, and so on. Thus, the carries ripple down through the digits, right-to-left, with the final leftmost digit sum being performed last. This approach is known as ripple carry addition.

Adders with "look-ahead carry" speed up addition by predicting carries and summing all digits at once.

Ripple carry addition can be speeded up by providing additional decide circuits to predict the carries for all digit positions at once. Fig. 3.18 shows the building block structure of such a 4-bit adder. This is called a look-ahead carry adder. With look-ahead carry addition, all digits are summed at the same time. No digit sum has to wait for the carry from any other digit sum. The result is a much faster addition. Typically, the adder of Fig. 3.18 can perform addition about four times as fast as the adder of Fig. 3.16. This is a good example of how providing extra decide circuits to our systems can often improve the system performance considerably.

Figure 3-18.
Full Adder with Fast Carry

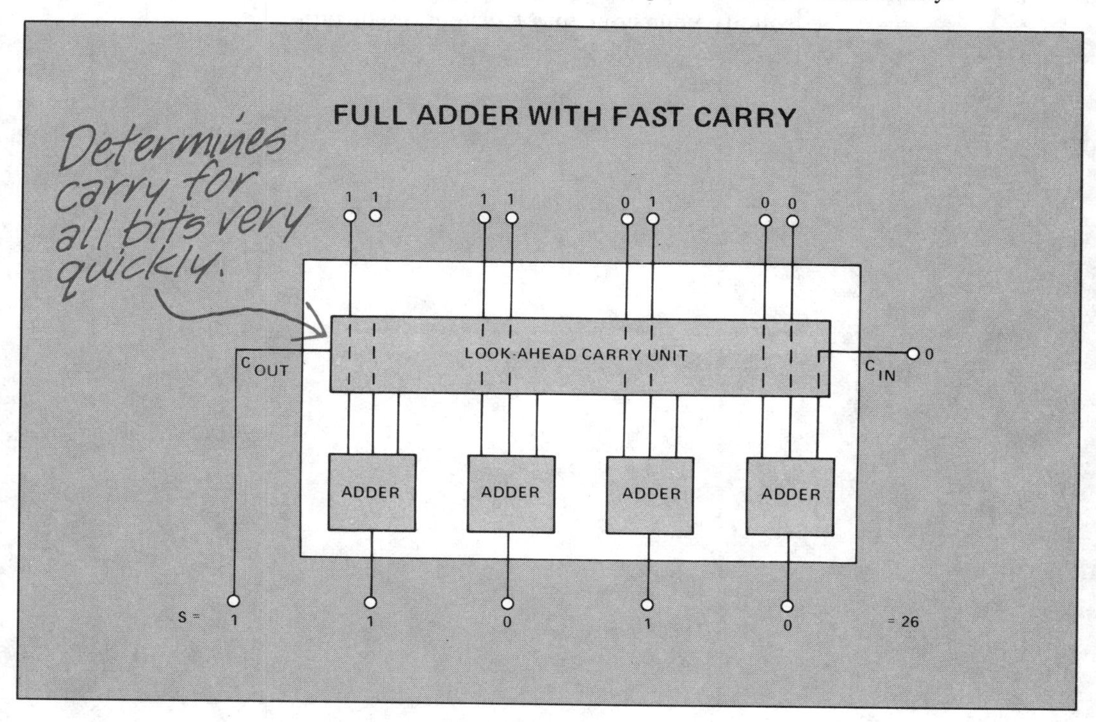

The 4-bit adder of Fig. 3.18 is available in a tiny single integrated circuit building block. TI calls it the SN74283 4-bit binary full adder. Such devices can easily be connected together to perform addition of 8-bit or 16-bit (or longer) binary numbers. Such combinations can perform the addition of two 16-bit binary numbers in less than 50 billionths of a second!

Now you can see why we have stressed the building-block concept. This single integrated circuit contains more than 300 transistors, resistors, and diodes on a single chip of silicon. Few people can keep even one circuit like that clearly in mind. Luckily, if you can master the building-block concept, you will have a perfectly clear idea of how a digital computer makes mathematical computations.

Of course, there is much more to digital computers than we have covered in this chapter. In a later chapter, we will discuss memory functions, which involve integrated circuits such as flip-flops, registers, and counters. But now that you know how the decide stage of a computer functions, you understand the working heart of the computer.

In the next chapter, we'll take the baby computer we've just built, and put it into a complete system. This exercise will help solidify your concept of system organization.

Quiz for Chapter 3

1. Strictly speaking, the circuits of which a system is composed:

- ☐ a. Sense and act, but do not decide
- ☐ b. Translate information from a sensing device into a form suitable for making a decision
- ☐ c. Add power to information from the decide section to drive "acting" devices
- ☐ d. Make decisions
- ☐ e. b, c, and d above

2. The two types of decision-making circuits, corresponding to the switching and regulating of electricity and to the two categories of code for transmitting information, are:

- ☐ a. Sensing and acting types
- ☐ b. Semiconductor circuits and vacuum-tube circuits
- ☐ c. Digital and analog
- ☐ d. Information and power
- ☐ e. The source and the point of use

3. Chapter 3 emphasizes one of the two categories of decision-making circuits because one of the most important groups of semiconductors is:

- ☐ a. Amplifier transistors
- ☐ b. Digital integrated circuits and switching transistors
- ☐ c. Analog integrated circuits
- ☐ d. Sensing-type circuits
- ☐ e. Acting-type circuits

4. The modulator in an amplitude-modulated (AM) radio transmitter is an example of:

- ☐ a. A digital decision
- ☐ b. How semiconductors replace vacuum tubes
- ☐ c. How amplification can be thought of as a decision-making process
- ☐ d. How to adapt a radio transmitter to make it useful as a multiplier circuit
- ☐ e. None of the above

5. Decisions in an electrical system:

- ☐ a. Always involve mathematical computations performed by circuits
- ☐ b. May involve not only numbers, but also physical variables like fluctuating air pressure, or even the statements "yes" and "no"
- ☐ c. Always require semiconductors
- ☐ d. All of the above
- ☐ e. None of the above

6. AND, OR, NOT, NAND and NOR refer to:

- ☐ a. Analog devices
- ☐ b. The logic gate: the simplest of digital decision-making circuits
- ☐ c. The parts of a thermostatic furnace control system
- ☐ d. Relays
- ☐ e. a and b above

7. The purpose of a truth table is:

- ☐ a. To be sealed inside a gate or memory
- ☐ b. To explain the internal construction of a digital logic circuit
- ☐ c. To help to analyze an analog circuit
- ☐ d. To prove which outputs are true and which are false
- ☐ e. To show in condensed form all possible combinations of inputs to a digital circuit and the resulting outputs

8. A digital system can make any logical or mathematical decision:

- ☐ a. By means of simple logic gates put together in large enough numbers and in the proper arrangement
- ☐ b. Provided it can be expressed in binary (yes-no) form
- ☐ c. By using only amplifier-type circuits
- ☐ d. If sufficient AND gates are used
- ☐ e. a and b above

9. Which are the most basic logic gates from which all digital logic circuits are built, including some other gates?

- ☐ a. NAND, NOR, NOT
- ☐ b. AND, OR
- ☐ c. AND, NAND
- ☐ d. OR, NOR
- ☐ e. AND, OR, NOT

10. The SN74283 binary 4-bit adder is:

- ☐ a. Implemented with relays acting as three kinds of logic gates
- ☐ b. A building block used in digital systems, made of smaller building blocks called one-bit full-adders and half-adders, which in turn are made of even smaller building blocks called logic gates, which are merely simple switching circuits
- ☐ c. A method of adding a pair of four-bit binary numbers to produce a five-bit binary sum
- ☐ d. Composed of more than 300 transistors, resistors, and diodes on a single tiny piece of silicon called an "integrated circuit"
- ☐ e. All the above except a

4 RELATING SEMICONDUCTORS TO SYSTEMS

Key Words

Capacitance
Flip-Flop
Inductance
Power Dissipation
Reactance
Shift Register

Definitions are found in the glossary in the back of the book.

Relating Semiconductors to Systems

In the last chapter, we developed a baby computer, and pointed out that this four-bit binary full-adder was contained in a single integrated circuit, the SN74283. Figure 4.1 shows that integrated circuit. We know now that it has the ability to make decisions and thereby solve simple mathematical problems. The answers it can provide are made with unerring accuracy, at fantastic speed. Yet by itself, it looks extremely helpless — and it is.

Figure 4-1. Integrated Circuit in 16-Pin Package

It has no way of receiving instructions from the outside world. How can it be told what numbers to add? How can it communicate its answers? How can it remember (or store) the input numbers long enough to add them? This integrated circuit is very much like a brain without a body. To function at all, this decision maker must be incorporated into a complete system. Let's take a look at what's required to make it work. Figure 4.2 presents a

Figure 4-2. Baby Computer to Perform Addition

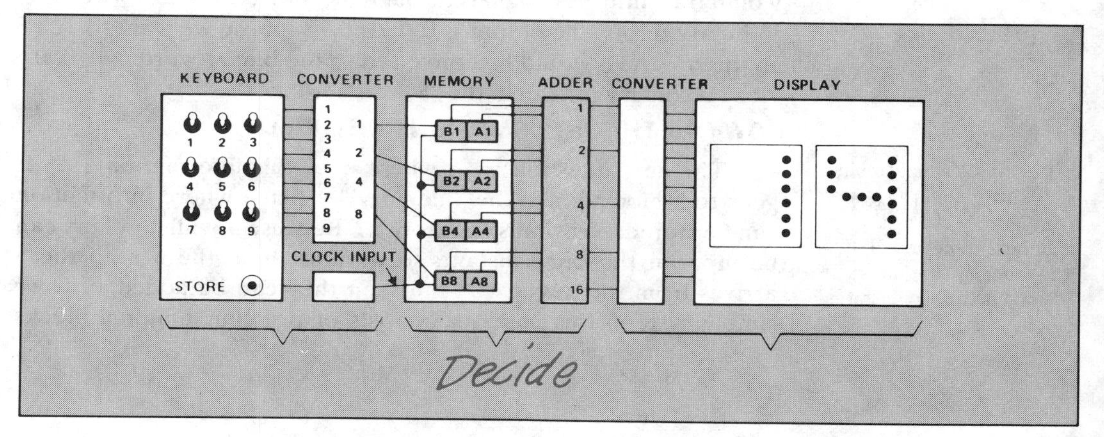

complete system incorporating the adder we described in detail in the last chapter. This is a very simple system, and all it can do is add two decimal digits. But the principles involved are the same as those behind a complex electronic calculator or computer. On the other hand, the figure may look very complex to you — so we'll break it down into its simple building blocks, one by one.

The box labeled "adder" is the integrated circuit we developed in the last lesson. It accepts binary numbers as inputs, adds them, and outputs the sum as a binary number.

HOW CAN THE SYSTEM BE TOLD WHAT NUMBERS TO ADD?

Numbers keyed into the example system are converted from decimal to binary format for addition. The converter circuitry, like the adder, is composed of logic gates.

Now let's start at the input end of the system and work our way through it. The first thing we need is something to sense the numbers to be added. So we use a keyboard, with a key for each of the digits 1 through 9. The keys are simply switches. Press one of the keys and you transmit a "yes" signal for the number chosen. Remember, however, that each of these keys is marked with a *decimal* number, and the keyboard senses your decimal command. But the *adder* uses *binary* numbers.

So the next thing the system must do is convert your decimal number to binary. Suppose you turn on the 6 key. This action produces a "yes" signal — that is, a high voltage — in the number 6 wire, which goes to the next block. This block is called a "decimal-to-binary converter." Like our adder, it is built of logic gates. It transforms the information it receives through the nine wires from the keyboard, into the binary form handled by the decide stage. Output from the converter is carried by four wires that transmit the binary version of the input numbers. As you will recall, the top wire stands for 1, the second wire stands for 2, the third for 4, and the last wire for 8. So when we input a 6 at the keyboard, we would transmit "yes" signals through the 2 wire and the 4 wire. Or, in binary terms, the output is 0, 1, 1, 0. Suppose we want to add 8 to this 6. The 8 would be converted to the binary word 1, 0, 0, 0. (It's customary to read in the "8-4-2-1" order.)

HOW CAN THIS INFORMATION BE STORED?

After the decimal-to-binary conversion, the system temporarily stores the numbers to be added in the memory section — also built from logic gates.

The next question is — where does this information go? It goes to a block called the "memory," a place where information can be stored. Why must we store it? Because not all numbers can be input at the same instant; so we must hold the 6 until the 8 arrives from the keyboard, and then they can be added. In our simple system, this memory consists of just four building blocks

The shift registers that compose the memory section are made up of flip-flops — logic circuits that hold or store information once they are set or reset to 1 or 0 according to the levels of the inputs.

called "shift registers." Notice that each register is divided into two parts or compartments. These two compartments are enough to store the two numbers to be added. Before the numbers are actually added, the four digits of our first number will be stored in the far right-hand compartments labeled "A." The second number to be added will be stored in the left-hand compartments labeled "B." But we can only put the numbers in one at a time.

Let's see how this storage is achieved. The shift registers do pretty much what the name implies — they register, or hold information, and they shift it. The two parts of the shift register, which we have called compartments, are actually switching circuits called "flip-flops."

There are several different kinds of flip-flops, and the particular type we have here is called "D-type." Flip-flops are used in digital systems for a very important function — that of *storing*, or *remembering*, data. Each flip-flop can store one bit of information — a 1 or a 0.

HOW DO FLIP-FLOPS WORK IN A SHIFT REGISTER?

"Clock" pulses synchronize the shifting of binary information through the flip-flops.

Each flip-flop in our system has an input and an output for digital information, plus a "clock" input which provides a control signal telling the flip-flop when to operate (See Fig. 4.3). The clock input is a binary voltage signal produced by the pushbutton labeled "store" on the keyboard. When the "store" button is pressed momentarily, it provides a "clock pulse" (the clock voltage goes high and then low again) to all the flip-flops. A clock pulse causes the output of a flip-flop to "flip" to a one or "flop" to a 0, depending on which bit is being presented at the input. When a D-type flip-flop receives a clock pulse, it "remembers" the bit that is being received at the input at that moment, storing this 1 or 0 and presenting it at the output until the next clock pulse arrives.

WHAT HAPPENS WHEN THE FLIP-FLOP FLIPS?

Figure 4.3 shows the succession of events in one of the shift registers as two numbers are stored. Consider the one shift register that stores information from the 4 wire. Suppose this wire is now carrying a logical 1. First, the clock pulse causes the output of the first flip-flop to flip to a logical 1 state. This condition continues — that is, the output remains a logical 1 — until a new input is entered on command of another clock pulse.

**Figure 4-3.
Storing Information
in Shift Register with
Flip-Flops**

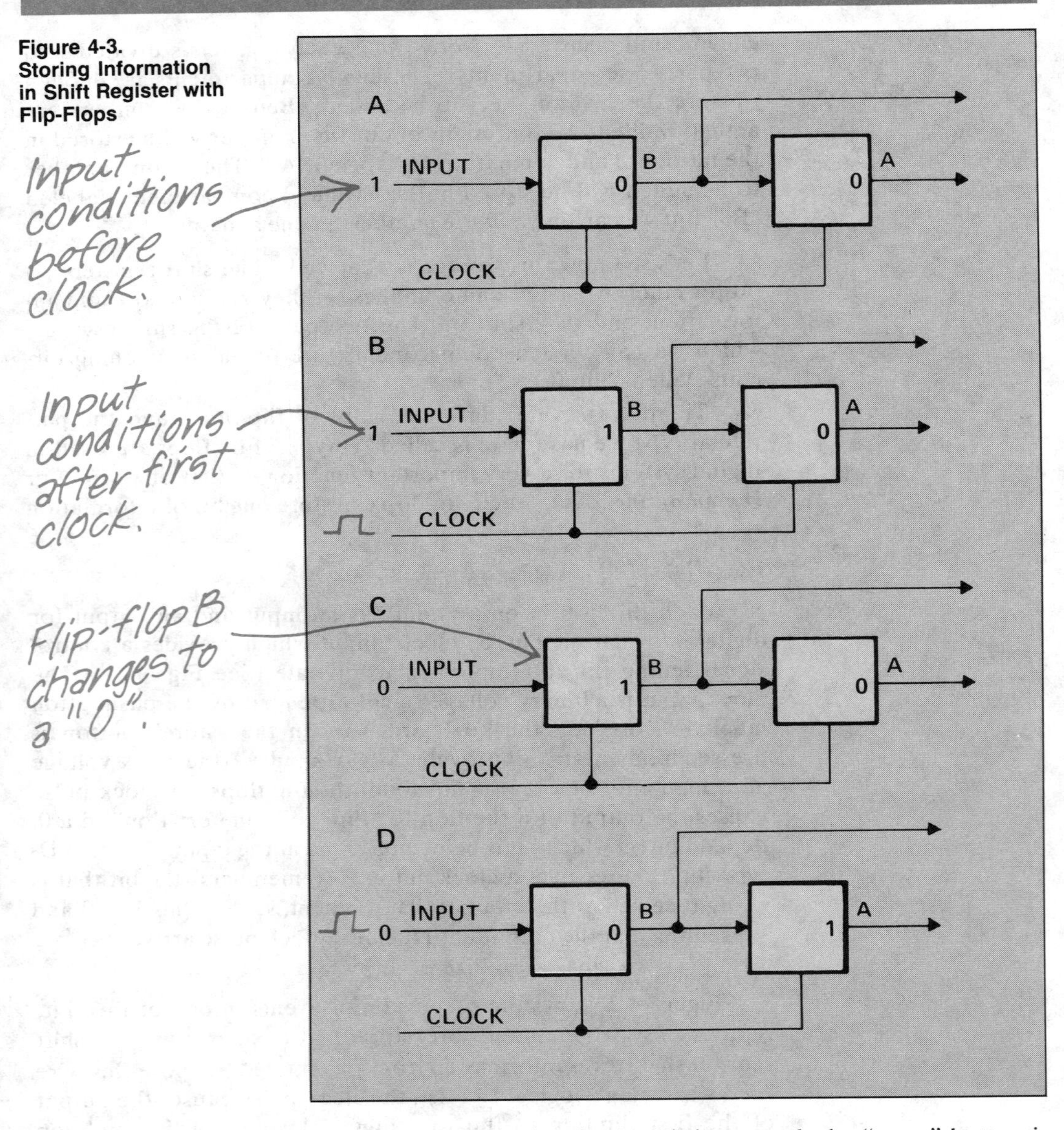

The new input is a 0. As this is entered, the "store" button is pushed, transmitting another clock pulse. This is where the shifting takes place. The input of the first flip-flop is 0. The input of the second flip-flop is 1, because that's what is being stored at the output of the first flip-flop. So on the signal of the clock pulse, the 1 is shifted to the output of the second flip-flop, and the 0 is shifted to the output of the first flip-flop. In this way, both desired digits are "stored" in their proper places.

There's nothing very mysterious about the flip-flops. They are simply switching circuits, put together in the right combination. The information in the memory is continually transmitted to the adder, which was explained in the last chapter; the adder decides what the sum is, and the answer appears almost instantaneously at its five outputs. But the information at the five *outputs* is in *binary* form, and we need a *decimal output.*

So the next thing we need to do is convert the 5-bit binary answer back to decimal form. So we employ another set of logic gates called, appropriately enough, a "binary-to-decimal converter." This in turn, drives the display devices.

HOW ARE THE DECISIONS DISPLAYED?

After the addition is complete, the system converts the sum from binary to decimal format for display. Like the rest of the system, the converter and display circuitry is built from logic gates.

The job of the act stage of this system is to display the answer from the decide stage output. There are many ways to display information, of course, so our system might drive printers to record answers on paper, or use gas-discharge tubes that make an illuminated display of the numbers. But in our sample system, let's say we're using light-emitting diodes — little semiconductor light bulbs — arranged in a matrix so that they can be selectively lighted to form digits. These are turned on at the right time by the output converter, and the answer appears on the face of the display — in our case, as 14, the sum of 6 and 8.

So in Fig. 4.2 we have a block diagram for a complete system built around the decision-making element we analyzed in the last chapter. As you see, it's a complete system with sense, decide, and act stages. It's admittedly a small system — a designer could probably put it all into a package the size of a pack of cigarettes — but now you can say you've analyzed a complete electronic system. And in this simple system, we have seen all the essential parts that go into even the world's largest computer — input, output, processing, and memory. This is about as far as we will go in our discussion of systems organization. Now, we're ready to see how semiconductors fit into systems.

Such factors as intended function, operating frequency, and power requirements determine the type of semiconductor devices used in a given system.

HOW DO SYSTEMS DIFFER FROM EACH OTHER?

We have pointed out the basic similarities shared by all electronic systems. But after all, systems are not all alike. It's true that all systems *can* be broken down into the stages of sense, decide, and act. And all systems *do* either manipulate information or do work. Nevertheless, systems differ from each other, and these differences have a distinct bearing on the kinds of semiconductors that go into them.

To illustrate these differences, compare our adding system of Fig. 4.2 with the radar system shown in Fig. 4.4. The operation of the radar is controlled by a central unit which we can call the decide stage. It responds to orders from the control console, which is a sensing function — in other words, an inputting of information. And we have the radar scope, a display of the information in the desired form. The act stage also contains a radio transmitter with a directional antenna, to transmit pulses of radio waves, on command from the decide stage.

You can quickly recognize the similarities between the two systems. But you can also recognize the differences between them. Let's spell out the differences. The most obvious difference is one we have already talked about: The adding machine uses digital information, and the radar, like most radio equipment, uses analog. And because of this difference in the way information is handled, you'll also recognize that the radar will contain chiefly amplifying circuits and amplifier-type semiconductors, whereas the adding machine employs switching circuits and switching-type

Figure 4-4. Radar System

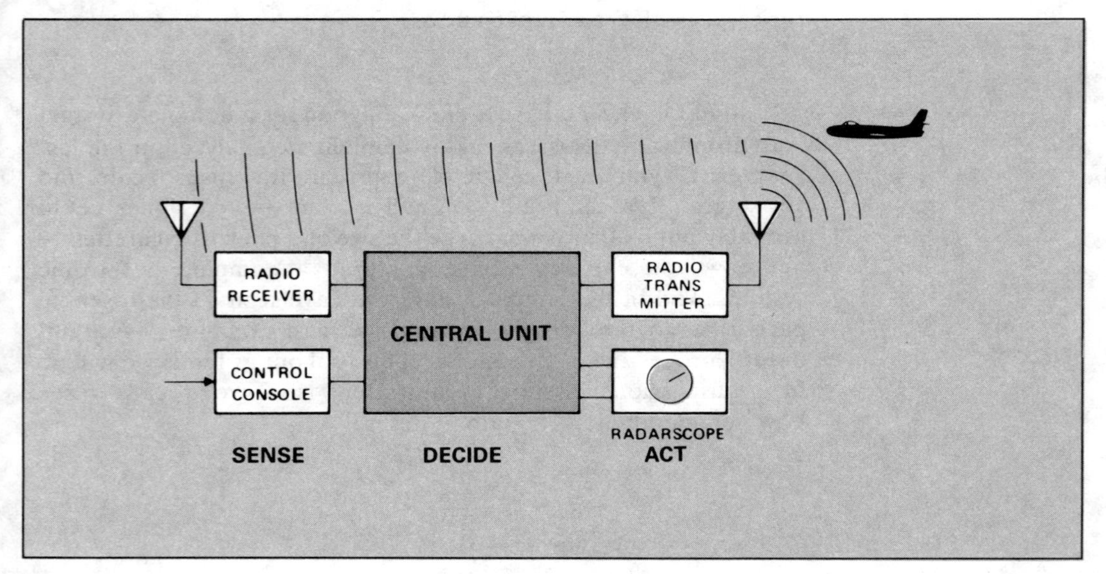

semiconductors. Moreover, radio transmission, as we have seen, depends on relatively high-frequency signals — whereas digital systems operate very well at comparatively low frequencies. The final major difference is that digital systems require just enough power to switch circuits; on the other hand, the radar system needs a strong radio pulse, one that will travel many miles and produce a strong enough reflection to return and be detected. In modern radar systems, this means high power semiconductors in the transmitter stage.

Let's summarize these differences:

Adding Machine	Radar System
Uses digital information	Uses analog information
Uses switching circuits and switching-type semiconductors	Uses amplifying circuits and amplifying-type semiconductors
Low frequency	High frequency
Low power	High power

Different sections of the same system often require different types of semiconductors.

We should also be aware that the differences we've listed here can exist *within* the same system. Consider the radar again. Although the radio receiver and transmitter use analog information, other parts of the system like the decide stage might employ digital information. Moreover, high-frequency circuits would be found only in the two radio sections. And not all parts of this system require high power; the decide section certainly doesn't, and the control console probably doesn't.

It's obvious, then, that the varying functions from system to system, and even within a system, require different kinds of semiconductors. And there are naturally, many detailed but important specifying requirements that lead a designer to choose one device rather another for each functions. Nevertheless, it is useful to divide all semiconductors into a certain few broad categories to clarify our thinking.

WHAT ARE THE BROAD CATEGORIES OF SEMICONDUCTORS?

Figure 4.5 presents a "family tree" covering all semiconductors. The basis for the categories is their application in

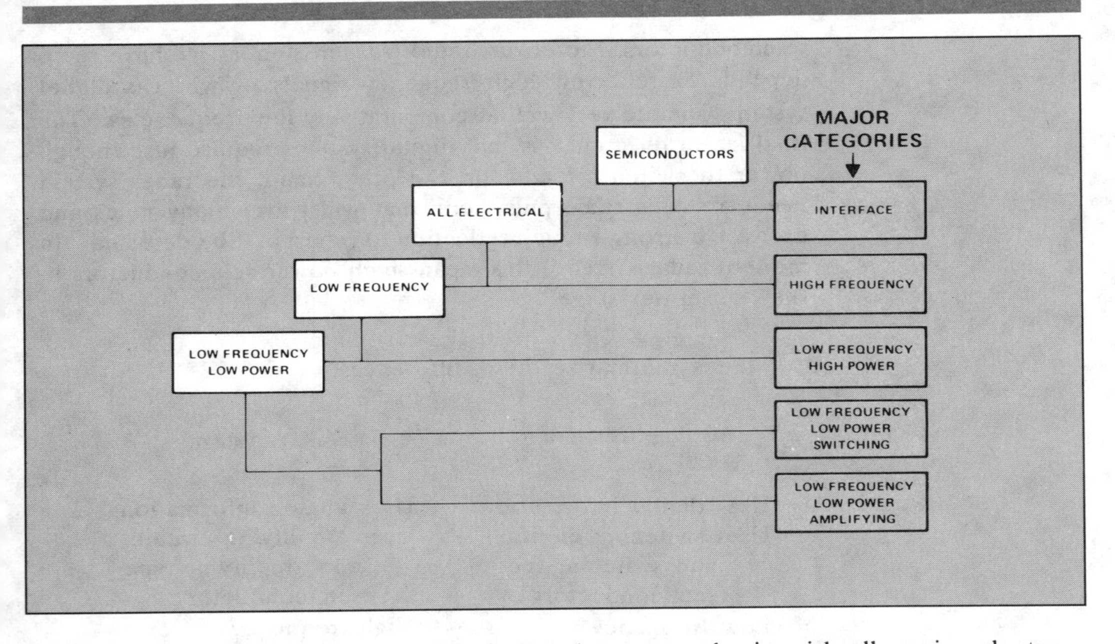

Figure 4-5. Semiconductor Categories

Semiconductor devices are classified according to their applications in electronic systems.

systems. Starting at the top, we begin with all semiconductors. They can be divided into "interface" devices and "all-electrical" devices. "Interface devices" is a category, but a relatively small one, made up chiefly (at this writing) of optoelectronic devices such as light sensors and light emitters. Interface devices are the only kinds of semiconductors that actually interface with the outside world — other semiconductors interface only with other electrical or electronic devices.

The rather loose term "all-electrical" embraces the vast majority of semiconductors. It divides into two branches. You'll recall that frequency was one of the major differences between the adding machine and the radar system. So we can establish "low frequency" and "high frequency" as sub-divisions.

Then from low frequency, we branch again, on the basis of power. So we have "low-frequency low-power" and "low-frequency high-power."

Finally, the low-frequency low-power semiconductors can be broken into two categories depending on whether they are switches or amplifiers. The end result of our family-tree breakdown is that we can place any transistor into one of the five broad categories represented in the *right-hand* column of the tree.

Before we go too far, bear in mind that the high-frequency category includes both high-power and low-power types, but regardless of power, all high-frequency circuits and devices have common characteristics that set them apart from low-frequency types. Similarly, bear in mind that the low-frequency high-power category includes both switching and amplifying circuits and devices.

You might like to refer back to Fig. 4.5 as we discuss typical devices in each of these categories, to entrench this framework firmly in your mind.

WHAT ARE SOME TYPICAL INTERFACE DEVICES?

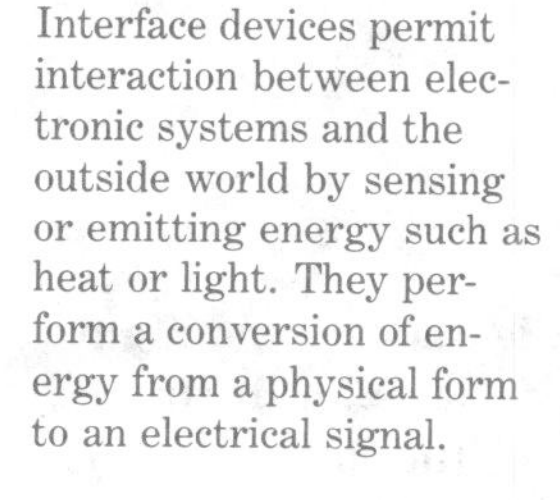

Interface devices permit interaction between electronic systems and the outside world by sensing or emitting energy such as heat or light. They perform a conversion of energy from a physical form to an electrical signal.

We can define interface devices as those that sense or produce external energy.

A typical interface device is the Sensistor® resistor. This device is not actually a junction device, but a special kind of resistor made of silicon. Its resistance changes with temperature; thus, it can sense the external energy of heat, and transmit this information electrically.

Optoelectronic devices are true semiconductors. Figure 4.6 is the electrical symbol for a phototransistor. Notice that it has an emitter wire and a collector wire just like a conventional transistor, but it has no electrical connection to the base region of the semiconductor element. Instead, external light energy passes through a little window or lens in the semiconductor enclosure and generates what amounts to a base current to turn the transistor on.

Figure 4-6. Phototransistor Symbol

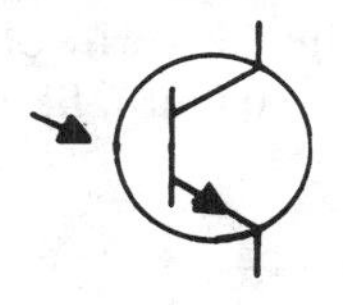

Figure 4.7 shows some typical light-emitting diodes (abbreviated LED). These function much like tiny light bulbs,

Figure 4-7. Light-Emitting Diodes

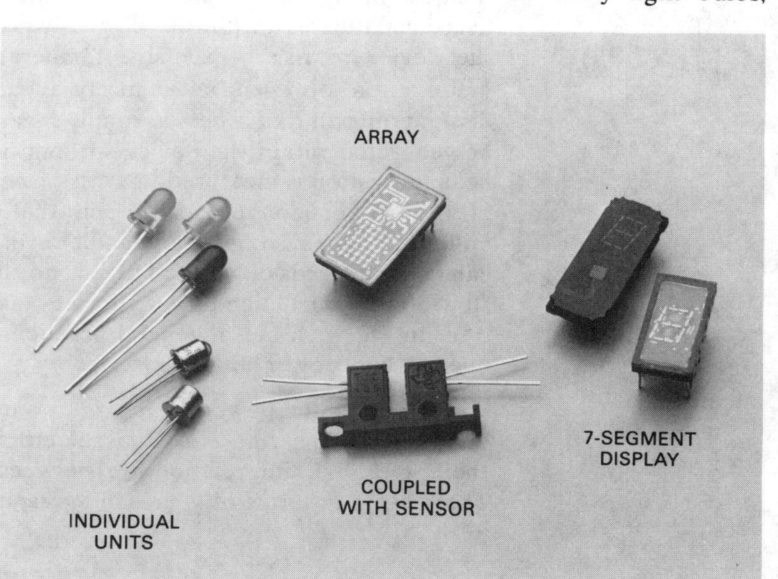

converting electrical energy into visible light. Light emitters are often used in arrays or matrices of the sort shown in Fig. 4.7, to display letters or numbers — you'll recall that we used such a display in our baby computer. Some of the advantages of LED's over other light sources are that they are very compact, they use little power, produce little heat, and work almost forever.

We'll be discussing optoelectronic devices at some length later on. For the moment, all we have to know is that light sensors convert light into electrical signals, and light emitters convert electrical energy into light.

HOW DOES POWER AFFECT CIRCUITS AND SEMICONDUCTORS?

One of the major characteristics of semiconductors and many other electric devices is the amount of power they can dissipate. In simple terms, "power dissipation" means the heat that is generated within a device by the "friction" of electrons rushing through it. Compare the device to your hand, and the electric current to a rope being pulled through your grasp. Your hand will be heated by friction — by which a certain part of the power in the moving rope is being dissipated (meaning wasted) in the form of heat. To dissipate power means to waste it through a friction-like process by conversion to heat.

"Power dissipation" in semiconductors is a measure of how much current-induced heat a device can tolerate without harm.

If heat is generated too rapidly within a device (that is, faster than the heat can be removed to the chassis or the air), then the heat accumulates within the device, causing the temperature of the device to rise. Excessive temperatures make the device malfunction or even fail entirely ("burn up"). The power-dissipation rating of a device simply means how rapidly heat can be generated within the device without harming it. This rate of heat generation is measured in watts or milliwatts, which are units of power. Semiconductors are built that can tolerate only a few milliwatts of heat, others can handle several watts, and still others can take hundreds of watts. There is no sharp borderline between "low-power" and "high-power" devices (as outlined in the "family tree" in Fig. 4.5), but in general practice, the line is set at about one watt of power dissipation.

Although the power *dissipated* by a device is not the same as the power being *transmitted* by electricity *through* the device, there is a definite relationship between these two quantities. Think again in terms of your hand grasping a moving rope. Two factors cause the heat — the speed of the rope (equivalent to electric current) and the pull you exert on the rope by grasping it tighter (equivalent to the voltage pressure exerted on the device — the difference between the voltages on both sides). The faster the rope and the tighter your grasp, the hotter your hand gets! Similarly, the greater the current and the greater the voltage drop, the greater the power dissipation. To put it simply: *amps times volts equals watts.*

Let's look at an example. Figure 4.8 is a simple motor-control circuit. We have a 12-volt power supply, an NPN amplifying

Figure 4-8. Transistor Circuit for Motor Control

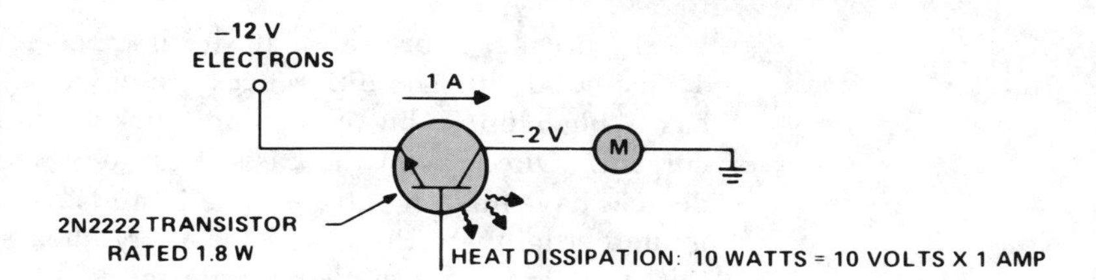

transistor, a motor, and a ground. The power supply provides 12 volts of electron pressure. Assume that we are withdrawing only enough electrons from the base of the transistor to allow one amp of electron current to flow from the emitter to the collector. Also assume that this is enough to maintain two volts in the wire to the motor.

Power dissipation in this transistor means the power that is wasted as heat in the process of regulation. Think of this heat as watts of power. We can compute the lost power by multiplying the voltage drop across the transistor by the amperage. Since we have 12 volts on one side of the transistor and 2 volts on the other, the voltage drop is 10 volts. Multiply one amp times 10 volts, and we get 10 watts.

Exceeding a device's power dissipation rating may damage or destroy the device.

But this transistor is rated at no more than 1.8 watts dissipation according to its catalog data sheet. What happens when we generate 10 watts inside it? All of this power must flow within a tiny chip of silicon, and the heat is generated in this chip. The smaller the chip, the hotter it will get. When the temperature of silicon rises above about 200 degrees Centigrade, or the temperature of germanium rises above about 100°C, the transistor goes out of control. It turns on all the way, and stays on. Soon, the transistor will be destroyed. In short, this transistor is a relatively low-power device (a small-signal device) that should never have been used in this circuit.

Large semiconductors with extra cooling features can dissipate more power than smaller devices.

So what kind of transistor do we need, to function properly in this circuit? First, we need a bigger chip of semiconductor material in the device. This will make the heat less concentrated, so the temperature will remain lower. Second, we need better contact between the chip and case. This will cut down heat insulation, so the heat can flow out of the chip more readily. And third, we need a bigger case to provide more area for the heat to be transferred to the surrounding air or to the plate on which the device is mounted.

Figure 4.9 shows a small-signal device, with a high-power device beside it. The differences are obvious. The power device has a big chip, a big case, and thicker leads to handle more current. Often, as in the case of the device shown here, power devices have built-in bolts — called "studs" — so that they may be mounted tightly to the chassis or a large heat sink. Even beyond this, they are often cooled by special air blowers or even water circulation.

The power device shown in Fig. 4.9 can dissipate up to 50 watts, more than enough to provide a wide margin of safety in our circuit.

You may be asking, "why not build all devices to handle high power? Then we wouldn't have to worry about malfunctions caused by heat." The first part of the answer is that they would be too big. Consider the binary full-adder we built for our baby computer. It has about 100 transistors in it — all very low-power devices. Imagine building this same function using big power transistors and circuits. Instead of a function as big as a cigarette pack, it would wind up being as big as a shoe box. Moreover, it would generate as much heat as an electric iron, so it would require some sort of special cooling system. Worse still, it would switch at very low speeds, so it wouldn't be practical for computer

Figure 4-9. Small-Signal Device Contrasted with High-Power Device

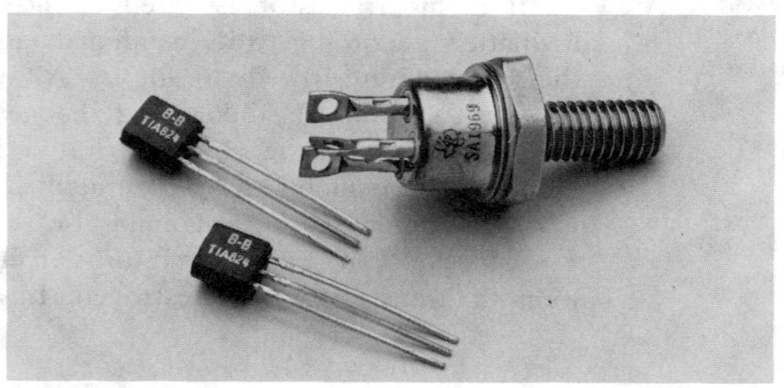

applications. Finally, it would be prohibitively expensive. The point we're making is that there are definite reasons for setting aside power semiconductors as a group.

Power semiconductors are distinguished not only by their appearance and power ratings, of course, but by their applications as well. They are found mainly in the "act" stage of systems, where they switch and regulate the power driving the working devices.

Before leaving power devices, we should point out that these distinguishing characteristics apply not only to transistors, but to other kinds of semiconductors, such as diodes and thyristors.

HOW DOES FREQUENCY AFFECT CIRCUITS AND SEMICONDUCTORS?

First of all, we need to go a little deeper into an understanding of frequency than we did in Chapter 2. Frequency can be defined as the rapidity at which an event occurs. In terms of mothers-in-law, you might be concerned with the number of times she comes to visit during the year. Perhaps the frequency of this event is four visits per year. In some cases that might be considered high frequency. A swimmer might do one hundred twenty strokes per minute, a higher frequency. And the wings of a bumblebee beat at a frequency of about two hundred cycles per second, a still higher frequency.

Frequency is the number of times that a cycle of a repetitive event occurs per second.

What about electricity? The most familiar example is the frequency of alternating current (ac). The frequency of alternating current is the number of times per second that the electrons make a complete cycle, moving first in one direction and then the other.

We can make the effect of this alternation visible by routing the current through an ammeter which displays the current on an oscilloscope tube. Figure 4.10 shows how one cycle of ac looks on the scope. As the current increases in the forward direction, the

Figure 4-10. Alternating-Current Wave

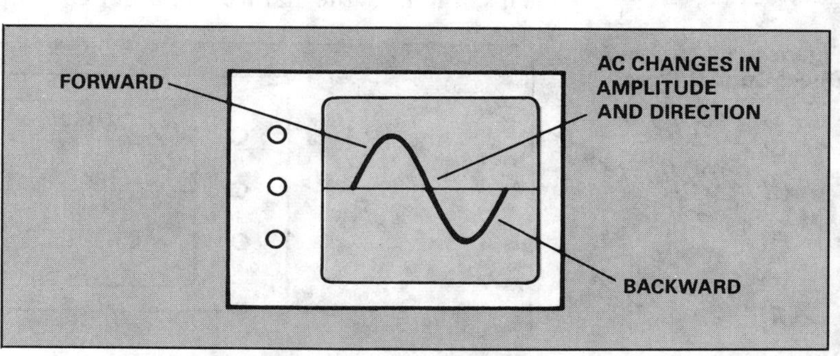

point of light on the tube indicates this by moving up. Then the light moves down as the current slows down, stops, and accelerates in the other direction. Then the current slows to a stop again. In the figure, the dashed line represents 0 current. The peak might represent plus 1 amp, and the valley might represent -1 amp — current in the opposite direction. The entire figure represents one *cycle* of alternating current. And remember that since current is driven by voltage, the voltage will also increase and decrease in a similar pattern.

In many cases, the frequency of alternating current in electronic circuits can be considered as waves of increasing and decreasing current, rather than as forward and backward motion.

We need to think of the frequency of alternating current not as a backward and forward motion, but rather as an increase or decrease in current. Now we can begin to see that fluctuations in current and voltage manifest themselves as waves. These waves "propagate" — that is, they travel by themselves once they are started in a wire — at a speed close to the speed of light.

You may know that semiconductors (except for the triac) don't operate with alternating currents. They use direct current. So why are we concerned with frequency and waves? We're concerned because waves can occur in *direct* current as well as in alternating current. We need to understand how we can have waves and frequencies without alternating current because it is the frequency of these waves that we are concerned with, when we speak of "high-frequency" and "low-frequency" semiconductors.

Let's refer again to the loudspeaker system we discussed earlier. In response to sound waves striking the microphone, the transistor causes the current from the power supply to the loudspeaker to vary — that is, to increase and decrease. The voltage supplied to the loudspeaker is also increasing and decreasing. If we routed this current through the ammeter and oscilloscope as we did with alternating current, we would get a similar picture of a wave, as in Fig. 4.11. The difference is that in this case, the lowest valley never falls below the zero line. That is, the current never reverses direction, but always flows forward. This, then, is the nature of direct-current waves.

Figure 4-11. Direct-Current Wave

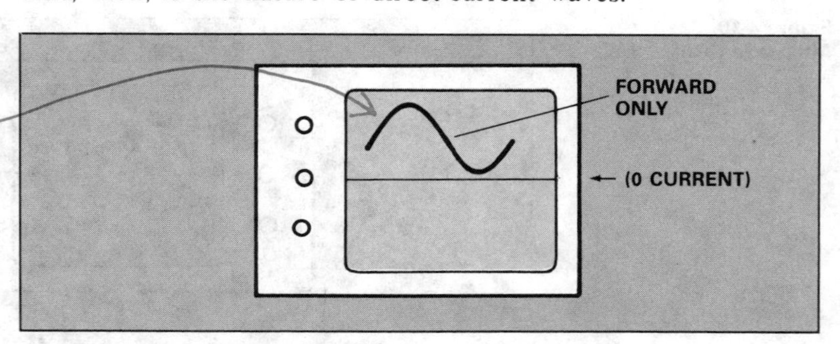

DC may change in amplitude but not direction.

Different types of systems operate at different frequencies.

The thing we are really concerned with is the *frequency* of these waves. Different systems, or parts of systems, operate at different frequencies. We pointed this out in our comparison of the digital adding machine with the radar system. Some circuits and some semiconductors operate well within a given frequency range, but will not perform properly at higher frequencies.

What keeps some semiconductors and circuits from operating at high frequencies? It's an aspect of the behavior of electrons called "reactance."

WHAT IS REACTANCE?

The "reactance" of electronic circuits — their resistance to changes in ac voltage and current — increases as operating frequency increases.

We have already seen how resistance impedes the flow of electricity. Reactance is the other factor that tends to impede it. But whereas resistance is always at work, reactance is present only when current or voltage are being *increased* or *decreased*. That is, reactance only impedes *changes* in voltage and current. The more changes we have in voltage and current — the higher the frequency — the more reactance we will get in the circuit. This is the basic reason why the performance of circuits and devices at high frequencies is different from that at low frequencies.

Visualize a simple circuit with a 12-volt power supply, a switching transistor, and a lamp. This is a special lamp that can be turned on and off as fast as we like. Suppose we are opening and closing the switch at a frequency of a million times per second. The dashed lines in Fig. 4.12 show what we want the voltage in the

Figure 4-12. Effects of Reactance on Voltage and Current Waveforms

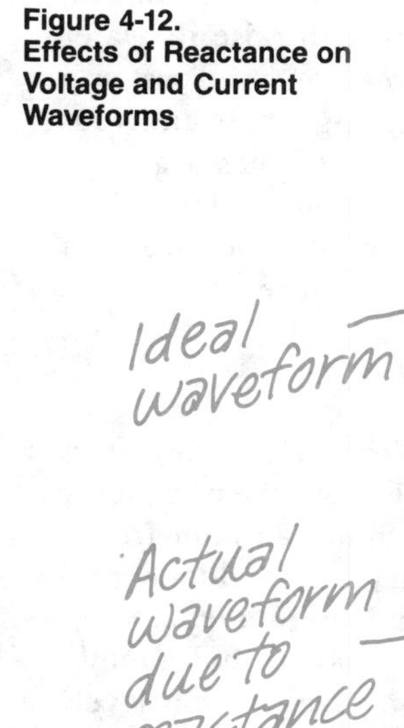

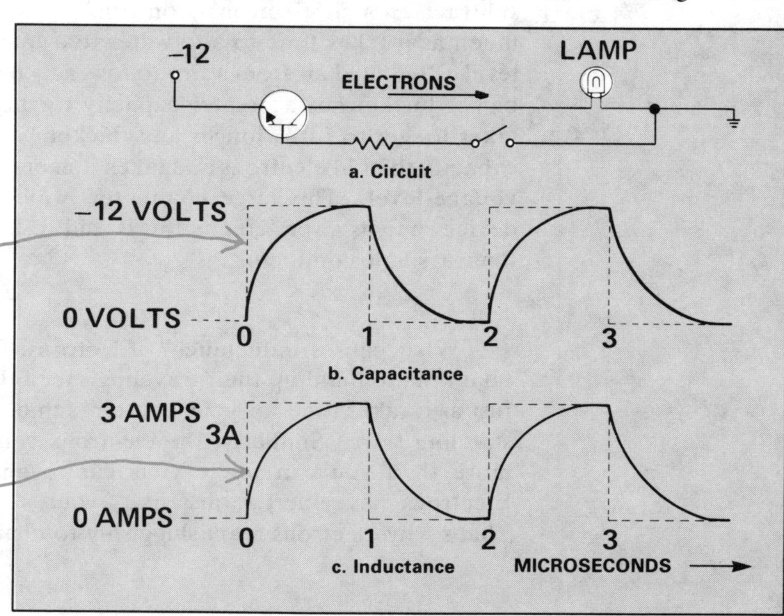

whole wire to do. The entire figure formed by the dashed line is called a "waveform." It says we want the voltage to rise instantly to 12 volts when we turn the switch on, then fall instantly to zero volts when we turn the switch off. When a millionth of a second (1 microsecond) has elapsed, the transistor is turned on again and the cycle is repeated. Similarly, we want the *current* to rise and fall just as sharply.

Unfortunately, these clean, sharp, square waveforms for voltage and current are ideals that can never be realized in actual practice. The villain of the piece is reactance. Reactance acts as a drag on any change of voltage or current, causing them to build up slowly over a period of time after the switch is closed, and causing them to drop off slowly after the switch is open. What we get in practice are the waveforms shown by the solid lines in the figure. And current waveforms look much like voltage waveforms — gradual rise, gradual fall.

Capacitive reactance opposes changes in voltage. Inductive reactance opposes changes in current.

The reactance characteristic that affects current is different from the one that affects voltage, but they are closely related. *"Capacitance"* is the dragging effect on changes in *voltage*. *"Inductance"* is the dragging effect on changes in *current*. Every part of a circuit suffers both of these forms of reactance.

What causes capacitance? Well, it takes time to add or subtract enough electrons from the wire to change its voltage, much as it takes time to add water to a bucket to raise its water level or to bleed air from a tire to lower its pressure. And just as a bigger bucket has a greater capacity for holding water and thus takes longer to fill, a longer and thicker conductor has a greater capacity to hold electrons and takes longer to fill up to the desired voltage level. This large conductor, which may be a wire or a device such as a semiconductor, is said to have more capacitance than a small conductor.

What causes inductance? Electrons, from a standing start, take time to build up their traveling speed through the conductor, just as it takes time for your car to reach 60 miles an hour from a standing start. Similarly, the electrons can't stop suddenly, any more than you can brake your car to an instantaneous stop. Electrons have inertia, just like your 4000-pound automobile. That's why electrons react sluggishly to changes in current levels.

HOW DOES REACTANCE AFFECT DIFFERENT FREQUENCIES OF OPERATION?

Now that we have some idea of the sluggish effects of capacitive reactance on voltage, and inductive reactance on current, the question is how do these properties affect circuits at high frequencies. Consider our lamp-switching circuit again, Fig. 4.13. Suppose we are operating our lamp at a low frequency — say 1 kilohertz. At this low frequency, turning it on and off 1000 times per second, the time drag caused by reactance is very short compared with the length of the pulse — so we get a waveform that is satisfactorily square. Our circuit has achieved its purpose, because the lamp is flashing at proper intervals. Capacitance and inductance are causing no problems at this relatively low frequency.

But what happens if we increase the frequency, say to 10 megahertz — 10 million pulses per second? In Fig. 4.13, the waveform marked "10 MHz" is what we want our voltage and current waveform to look like. Here again, we want sharp, square pulses, with instantaneous starts and stops as the transistor is switched on and off. But because of reactance, we get the stunted solid curve instead. The voltage and current hardly have time to get going before the transistor switches off and they are knocked

Figure 4-13. Effects of Reactance At Different Frequencies

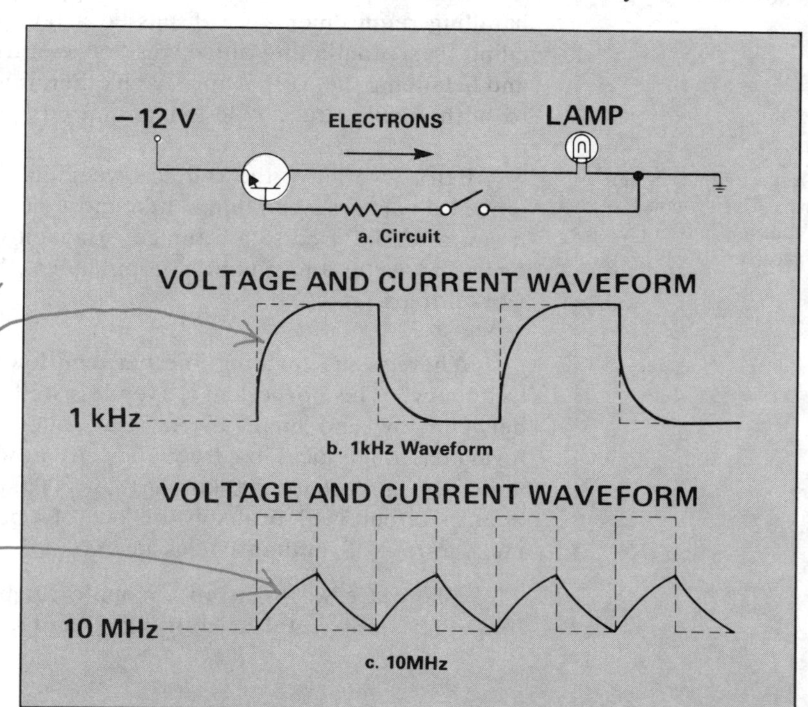

Waveform satisfactory at low frequency.

Effects of reactance at high frequency.

down again. Then they may not even reach zero before the switch hits them with power and they are turned on and start up again. Obviously, performance of the circuit is nothing like what we require. So we say that this circuit has too much reactance to handle a frequency this high. The lamp doesn't turn on and off cleanly at the desired intervals — it just sits and flickers at us.

The situation would be even worse if we were driving a transistor instead of a lamp. If the top of our waveform represented the threshold voltage level for turning a transistor on, our high-reactance circuit operating at this frequency would never build up enough input voltage to switch the transistor.

The use of small components with short connections reduces the effects of undesired reactance at high frequencies.

Now you can see what we have been leading up to. Circuits operating at high frequencies have many special design considerations, both for the devices and the circuits. One very basic requirement is for very small semiconductors and other devices, interconnected by very short lengths of small-diameter wire — because the less conductor material involved, the smaller the reactance.

It has probably already occurred to you that these high-frequency requirements are just the opposite of the good power-handling requirements we discussed a few pages back. Reconciling these conflicting objectives between frequency and power, and balancing them off against each other, is a continual challenge to both semiconductor designers and equipment designers.

In some circuit designs, capacitive or inductive reactance is a desirable characteristic.

Before we leave the subject, we should point out that reactance isn't always a bad thing. Like most electrical characteristics, it can often be used to advantage. Capacitors and inductors are two very useful and common components that put reactance to work for us.

Where is the dividing line between low frequency and high frequency? This borderline is even less well-defined than the one between low and high power. But roughly speaking, the performance of a typical low-frequency circuit will have deteriorated noticeably at perhaps 300 kilohertz — 300,000 cycles per second. Such a circuit will probably be out of operation entirely at 3 megahertz — 3 million cycles per second.

Where are we likely to encounter high frequencies? High frequencies are useful in both digital and analog equipment. In digital computers, the high frequencies of switching circuits — presently as high as 200 megahertz — make possible the incredibly fast operation and amazing "number-crunching" capabilities of modern computers. Analog equipment using amplifying-type circuits use even higher frequencies; in communications, radio waves are transmitted and received over a very broad spectrum of frequencies — ranging up to hundreds of gigahertz — billions of cycles per second.

The various types of digital and analog equipment operate over a wide range of frequencies, ranging from thousands to billions of cycles per second.

If you're familiar with the radio broadcast bands, you know how broad a spectrum this is. Very low-frequency (VLF) radio waves from 10 to 30 kilohertz, for example, are used in transmission to submerged submarines. Low frequencies are used by surface ships, among others. Our most familiar radio equipment falls into the medium-frequency category, used for AM broadcasting, and aircraft and mobile uses. High frequencies are used for short-wave broadcasting over long distances. Very high frequences (VHF) and ultra-high frequencies (UHF) are used for FM broadcasting and television. Even higher than these are the regions called "microwave" frequencies, which are used for straight-line radio communication and radar. The next step beyond these frequencies — is infrared light!

Before taking the quiz for this chapter and then pressing on, you might like to take a moment to review the semiconductor "family tree" shown in Fig. 4.5, just to be sure you have the characteristics of the major categories clearly in mind.

At this point, we have learned enough about the use of semiconductors in circuits and systems for our purposes. In the next chapter, we begin to examine the semiconductor devices themselves, starting with the simplest of all, diodes. In subsequent lessons, we will proceed through transistors, thyristors, optoelectronic devices, integrated circuits and others.

Quiz for Chapter 4

1. In digital gates and flip-flops, a higher voltage level usually indicates:

☐ a. "1"
☐ b. Yes
☐ c. "0"
☐ d. No
☐ e. a and b above
☐ f. c and d above

2. In shift registers made up of several flip-flops, the clock signal:

☐ a. Tells what time it is
☐ b. Indicates when to shift a bit of data from the input of the flip-flop to the output
☐ c. Is a bit of information stored in the flip-flop
☐ d. None of the above

3. Reading from **right to left,** binary digits represent 1, 2, 4, 8, 16, 32, etc. Thus, the decimal number "10" is written in binary number code as:

bit values: 8 4 2 1

☐ a. 1 0 1 0
☐ b. 0 0 1 0
☐ c. 1 1 1 0
☐ d. 1 1 1 1
☐ e. None of the above

4. The section of a simple digital adding machine which makes significant decisions but does not store information is:

☐ a. Memory
☐ b. Adder
☐ c. Input keyboard
☐ d. Decimal-to-binary converter
☐ e. Binary-to-decimal converter
☐ f. Output display

5. Designing semiconductors which can carry large currents involves:

☐ a. Increasing the size of the semiconductor chip
☐ b. Improving the degree of contact between the chip and case
☐ c. Making the case bigger so that it transfers more heat to the surrounding atmosphere and chassis
☐ d. None of the above
☐ e. a, b and c above

6. Semiconductors considered to be "low power" or "small signal" usually have power-dissipation ratings of:

- ☐ a. 20 watts or less
- ☐ b. 10 watts or less
- ☐ c. 5 watts or less
- ☐ d. 1 watt or less

7. Reactance is a property of electrical devices and conductors which:

- ☐ a. Resists the flow of electrons
- ☐ b. Retards changes in voltage
- ☐ c. Retards changes in current
- ☐ d. a and b above
- ☐ e. b and c above

8. Some ways to help semiconductors perform better at higher frequencies are:

- ☐ a. Decreasing the size of the semiconductor chip
- ☐ b. Making the leads shorter and smaller
- ☐ c. Using smaller packages
- ☐ d. All of the above
- ☐ e. None of the above

9. Waves in direct current can involve:

- ☐ a. Regular changes in the direction of electron flow
- ☐ b. Regular changes in voltage levels
- ☐ c. Regular changes in current levels
- ☐ d. Frequency, just as with waves of alternating current
- ☐ e. a, b, and c, above
- ☐ f. b, c, and d, above

10. Compared to small-signal transistors, high-power transistors usually:

- ☐ a. Cost more
- ☐ b. Make more heat
- ☐ c. Are bigger
- ☐ d. Are more complex
- ☐ e. Switch slower
- ☐ f. All of above
- ☐ g. All but d and f above

DIODES: WHAT THEY DO AND HOW THEY WORK

Key Words

Capacitor
Clamping
Detection
Electric Charge
Electron
N-type Semiconductor Material
P-N Junction
P-N Junction Diode
P-type Semiconductor Material
Proton
Rectification
Semiconductor Material

Definitions are found in the glossary in the back of the book.

Diodes: What They Do and How They Work

Diodes are the logical starting point in our discussion of semiconductor devices for two reasons. First, diodes are the simplest kind of semiconductor. And second, the basic understanding of semiconductors we achieve from studying diodes is applicable to other types of semiconductors, including transistors, integrated circuits, and even large-scale integrated circuits (LSI). Up to this point we have dealt almost exclusively with what semiconductors *do*. We approached the subject this way because it is difficult to understand semiconductors unless you know what they're designed to accomplish.

WHAT ARE THE SIGNIFICANT EXTERNAL CHARACTERISTICS OF A DIODE?

Solid state diodes have two leads. "Diode" means an active device with two terminals.

The name "diode" simply means "having two electrodes." A diode is simply a package with two leads, or wires. The only significant external characteristic of a diode is its size. The small device in Fig. 5.1 is a diode. The bigger one may be more familiar to you by the name "rectifier," but a rectifier is just an overgrown diode. Now that you know something about power dissipation, you've probably concluded that the smaller devices handle less power than the big ones, but usually at a higher frequency. And recalling the trade-off, you can conclude that large diodes (rectifiers), with their heavy heat-dissipating leads, can handle more power, but at lower frequencies. This is generally correct.

Figure 5-1. Diodes

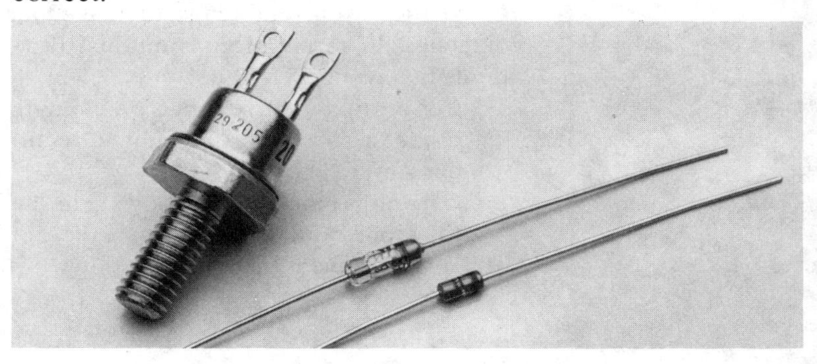

WHAT DOES A DIODE DO?

Diodes permit current flow in one direction only.

The most important function of a diode is to act as one-way valve for the passage of electrons. The diode permits electrons to flow through it in one direction, but bars their passage in the other direction. Therefore, a diode is basically a switching device rather than a regulating device.

The schematic symbol for a diode is shown in Fig. 5.2. The direction in which electrons can pass is *opposite* the direction in which the arrowhead points. Electrons flow within the diode from cathode to anode.

Figure 5-2. Current Flow Through a Diode

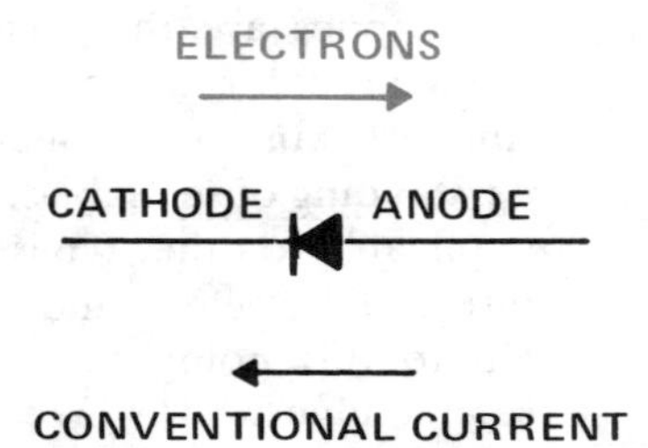

Due to tradition, most devices and circuit descriptions use a direction of current flow opposite to that of electron flow—an imaginary "conventional current" that flows in a positive-to-negative direction (*with* the arrows in transistor and diode symbols and *opposite* the direction of electron flow).

You may wonder why the diode symbol arrow points opposite to the direction of electron flow. In circuits, we like to use a current flow opposite to electron flow. Electrons flow from more negative voltages to more positive voltages. Circuit descriptions use an imaginary current that flows from more positive voltages to more negative voltages. This imaginary current flow is called conventional current. The diode symbol points in the direction of conventional current flow, since it is a circuit symbol. The conventional flow is the direction current would flow if the flowing particle were positively charged. Physically we see that current is due to negatively charged electrons flowing in one direction. This is equivalent to positively charged particles flowing in the opposite direction. In physical descriptions of devices, we tend to use the physically meaningful electron flow. In circuit descriptions, the conventional current tends to make our circuits simpler to understand. In terms of conventional flow, we can summarize our diode behaviour as follows:

1) Conventional current can flow only in the direction of the diode arrow. This current direction is called forward current.
2) Current is blocked from flowing in the direction opposite to the diode symbol arrow. Thus, reverse current is negligible.
3) When conventional current flows in the forward direction

the cathode voltage is the same (possibly, slightly less positive) as the anode voltage.

We will go into the difference between conventional current and electron current in more detail in Chapter 6. In the meantime, we will use conventional current in the circuits of this chapter.

WHAT IS A TYPICAL RECTIFIER APPLICATION OF A DIODE?

A primary application of diodes is to convert alternating current into direct current. Figure 5.3 shows such an application and also shows what happens to the current waveforms. The power supply is an ac generator driving a dc motor. Waveform "A" is the normal ac you're already acquainted with; electrons starting from a standstill, accelerating in the forward direction, slowing down to a stop, then accelerating to a maximum in the reverse direction. This is fine if we're running an ac motor. But motors are finicky, and this dc motor will only accept dc.

Rectifiers are diodes used to convert alternating current to direct current.

Visualize what happens when we put a diode in the circuit. It will pass the top half of the wave form, but not the bottom half. It's as though we erased everything below the dashed line. All of the pulses of current that get through are headed in the same direction. This function of the diode is called "rectification." That's why diodes built for this specific purpose are called "rectifiers".

**Figure 5-3.
Using a Diode as a Rectifier**

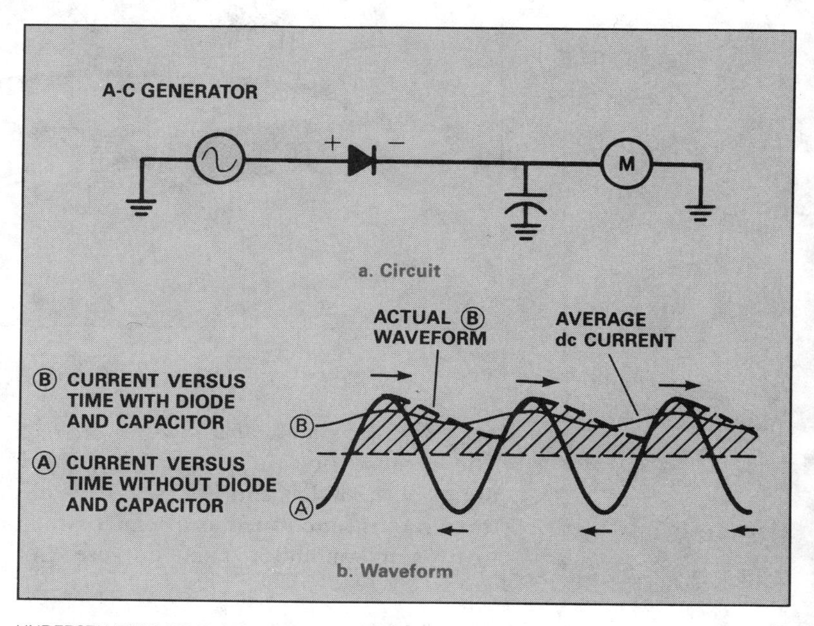

Capacitors in rectifier circuits complete the ac-to-dc conversion by smoothing the ripples in the diode output.

The current pulses we have left after rectification don't provide a very smooth source of power, unfortunately. So this circuit gives us a chance to demonstrate what a capacitor can do in a circuit. When we place the capacitor between the generator and the motor, the result is to smooth out the waveform as shown in "B." It's a lot like cutting a road through a series of hills, cutting through the peaks and using this earth to fill in the valleys. How does the capacitor do this? Remember, capacitance is the effect which retards sudden changes in voltage. A capacitor does this by acting as a storage reservoir for electrons. Imagine a large water tank, as in Fig. 5.4. The little man is repeatedly dumping buckets of water into the tank. So the water is being dumped into the tank in spurts, or pulses. But on the other side of the tank, the water will flow out in a fairly constant, smooth stream. Capacitors store and release electricity in much the same way.

Capacitors and capacitance act to resist or retard sudden changes in voltage.

Since this particular diode is driving a motor, it must be a high-power device, much like the one we saw in the photo. It doesn't have to switch at high frequencies — just 60 Hertz (cycles per second) like the house current it handles.

Figure 5-4. Capacitor-Water Tank Analogy

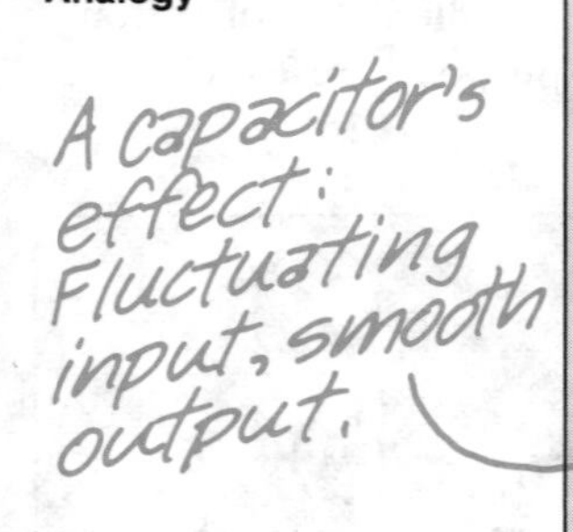

WHAT IS A TYPICAL LOW-POWER APPLICATION OF A DIODE?

Now, let's look at a low-power, or small-signal, application. The earliest diode application can be found in the old crystal radios of the 1920's and 1930's. The lead-sulfide (galena) crystal in these radios functioned as a semiconductor diode, long before the word "semiconductor" came into use. Let's see what it does.

A diode used to separate information from radio signals is called a detector.

Figure 5.5 presents an AM radio transmitting and receiving system. Below each stage of the system is a drawing of how the waves look at that stage. We talked about these waves earlier. You'll recall that, in radio transmission, a low-frequency microphone signal is used to modulate the amplitude of high-frequency waves produced by an oscillator. The modulated signal is converted to radio waves by the broadcast antenna, and then converted back into electrical waves by the receiving antenna. The crystal diode, one with moderate power and frequency ratings, cuts off one side of those modulated waves, just like the rectifier did for the motor in our previous example. This rectified signal drives the earphones. The earphones "average out" the radio-frequency pulses much as a capacitor does. The earphones are simply unable to respond to every little pulse of the high-frequency carrier waves, so they average out the current — and in so doing, provide sound which is a quite accurate duplication of the microphone signal.

The microphone is said to "modulate" the signal, because it modifies the amplitude of high-frequency waves from the oscillator. The reverse process, performed by the diode, is therefore called "demodulation," or "detecting." Therefore, a diode employed for this purpose is called a "detector." Modern semiconductor diodes do the same thing as the old galena crystal, but they're smaller, far more reliable, handle more power, and are less expensive. Since receiving ordinary AM broadcasts and driving earphones requires only a moderate power rating and a moderate frequency rating, diodes used for this purpose are typically called "general-purpose" diodes.

Figure 5-5. Transmitting and Receiving AM Radio Signals

Oscillator generates high frequency carrier.

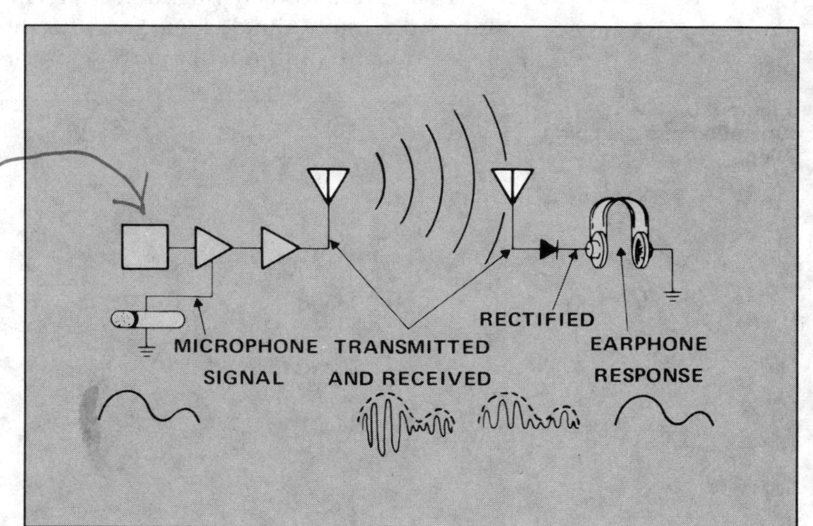

WHAT IS A CLAMPING DIODE?

Diode clamps protect sensitive circuit components from surges of high voltage.

"Clamping" means preventing voltage in one wire from exceeding the voltage in a second wire. Let's look at just one example of the many applications of clamping. In the circuit of Fig. 5.6, a diode clamp is used as a sort of safety valve to protect the transistor from damage due to surges of high voltage. The transistor is being used to turn current on and off through an electromagnet coil. This coil might be part of a motor or relay; like all coils, it has a great deal of inductance. Once the electrons in the coil are moving, they don't want to stop even after the transistor is turned off. When the transistor is turned off, the current is blocked at point "A." This is a little like shutting the corral gate against a herd of wild horses. The electrons keep coming, and pile up to a high voltage. If they build up too high, the current could break through the transistor at a dangerously high voltage, and damage or destroy it.

The solution is simple. We insert a diode across the coil as shown. The diode won't allow electrons from the power supply to detour around the coil and into the transistor. But it *will* allow electrons to bypass the coil in the other direction. So if the voltage at "A" goes above the power supply voltage (5 volts), then the diode will allow electrons to pass, draining the excessive buildup of electrons away from the transistor. The collector of the transistor is therefore "clamped" at 5 volts. The voltage can be below 5 volts, but will never be very much above it.

In addition to rectifying, detecting, and clamping, diodes are also used to build simple logic gates for digital computers. More diodes are used for this latter purpose than for any other.

Figure 5-6. Operation of a Clamping Diode

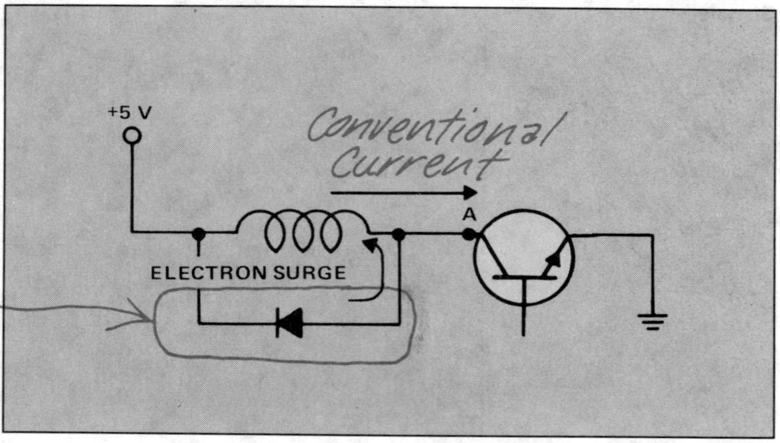

HOW ARE DIODES USED TO BUILD LOGIC GATES?

Figure 5.7 shows a diode-built two-input OR gate. Remember that the OR gate output is "one" (one means a high voltage) only if

Figure 5-7.
An OR Gate Made Using Diodes

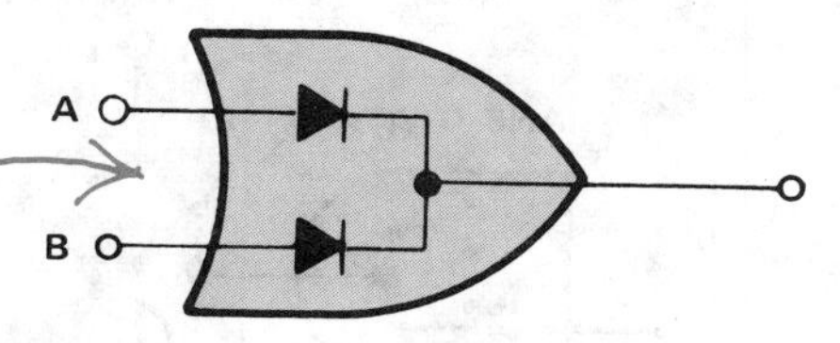

Diodes keep current from entering at one input and leaving through the other.

there is a "one" at input "A" OR "B." An input of "one" would clamp the output voltage at a "one."

Conventional current at the inputs can move only *into* the inputs, and not out of them, because of the one-way-valve quality of the diodes. Assume a 1 at input "A," and a 0 at input "B." Current would flow from the high voltage at the input to produce a high voltage (a 1) at the output. The only way to get a 0 at the output is to have 0's at both inputs. It's so simple that we have to ask the question "why do we need the diodes at all — why not just let the electrons flow through wires?" To answer that, consider what would happen if we had a 1 at input "A" and a 0 at "B." Without the diode on the 0 input path to block outgoing current, we would have a short-circuit path, and current would run out through this path, rather than out the output. The output voltage would then lie at some indefinite point between high and low. The output would not be a clear-cut yes or no — it would be a "maybe." Computers hate "maybes," so we *need* diodes to build this kind of simple logic gate.

We're far enough along now that you can begin to explain some things to yourself, so try this: If we turn the diodes around so that they block current from coming into the inputs, we would have an AND gate. Can you see how this works? We'll show you, in a moment.

In conjunction with power-sustaining devices, diodes can be used to build logic circuits.

A diode gate, like any electric circuit, uses power but *does not add* any power to make up for the loss. So if we were to string thousands of these gates together as required in a large computer, we would have to supply tremendous power to the input of the first gate in order to get any signal out of the last gate. This is quite impractical. So diode gates, while they have wide applications in digital systems, cannot be used exclusively. They must be associated with circuits that provide power.

One typical solution to this problem is to *boost* the power of the output of the diode gate by means of a transistor. Figure 5.8 illustrates this solution. In this case, we see the diodes performing

Figure 5-8. NAND Gate Made Using Diodes and a Transistor

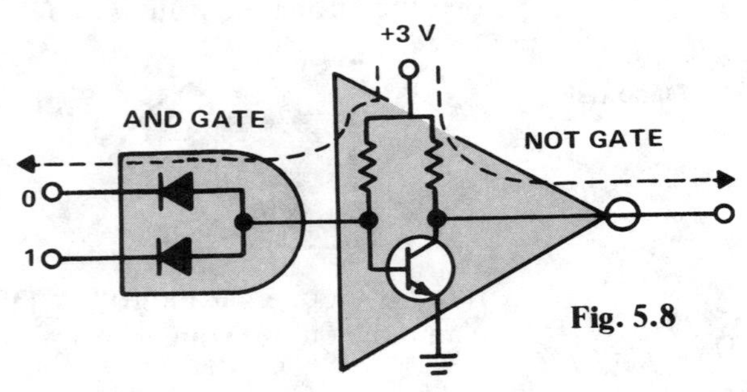

an AND function and making up an AND gate. The output from this gate goes into a switching transistor amplifier that adds power to the signal, since it has a power supply and ground connection. This switching transistor also acts as an inverter (NOT gate). A low voltage applied to either input will clamp the transistor base at ground, turning it off to supply a high output voltage.

The technique of teaming diodes with transistors to form logic circuits such as NAND gates is called "diode-transistor logic."

When we put an AND gate and a NOT gate together, they are usually thought of as one building block, a NAND gate. The symbol for a NAND gate is shown in Fig. 5.9. You can see where this has led us. We have built a logic gate that employs both diodes and transistors. This is diode-transistor logic, called "DTL." Such building blocks are built in integrated-circuit form

Figure 5-9. NAND Gate Symbol

NAND GATE

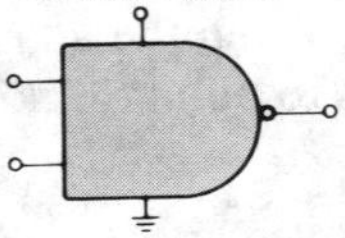

for modern computers, and rarely use discrete devices anymore. We'll discuss integrated circuits later on. For now, it's enough that you understand the function of diodes in logic circuits — it's the same, whether the diodes are in their own packages, or are built on the same chip of silicon with thousands of other devices. The diode applications we have covered — power rectification, detecting, clamping, and diode logic — account for perhaps 95 per cent of all uses of diodes. And they all make use of the one-way-valve property of the diode.

The principles of diode operation are common to all semiconductor devices.

Now that we have seen some of the uses of diodes, and discovered what the diode does, we want to find out how a diode works. Once you understand the basic properties of the diode, it's easy to understand all other semiconductors — because the principles underlying diode operation apply to all solid-state devices.

WHAT PROPERTIES OF SEMICONDUCTOR MATERIAL ARE IMPORTANT TO DIODE CONSTRUCTION?

Atoms of silicon, the most popular semiconductor material, consist of a nucleus of positive particles called protons (and other neutral particles) surrounded by negatively charged particles called electrons circulating in orbits around the nucleus.

First, we have to consider the properties of the diode's basic material, a silicon crystal. We'll talk about silicon because it is by far the most popular semiconductor material. But bear in mind that germanium and all other semiconductor materials follow the same general principles.

Like all matter, silicon is made up of atoms. At the center of the silicon atom, as shown in Fig. 5.10, is a concentrated mass

Figure 5-10. Silicon Atom

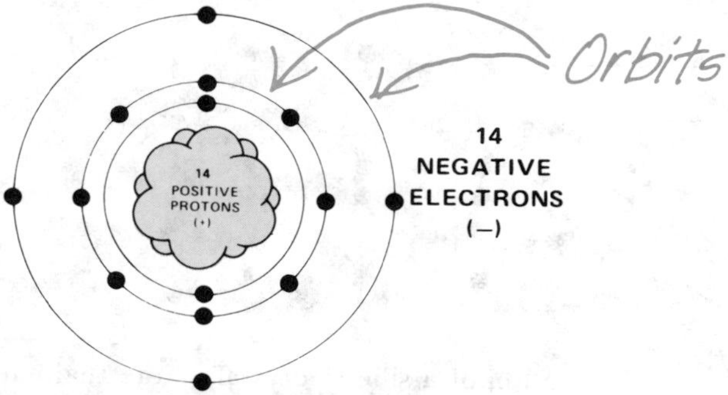

called the "nucleus." The nucleus contains fourteen electrically charged particles called protons, plus some neutral particles which we can ignore. Circling this nucleus like little satellites are fourteen other electrically charged particles called "electrons." When we say "electrical charge" we mean simply that a proton and an electron attract each other. Two electrons, on the other hand, repel each other. Moreover, two protons repel each other. This very basic rule of nature is expressed: *Like particles repel, unlike particles attract.*

The positive charge on protons and the negative charge on electrons are opposite and equal.

Proton and electron charges are not only opposite, they are equal. This means that a proton and electron together are electrically neutral — the equal unlike charges neutralize each other. So this combination neither attracts nor repels any other particles. Instead of referring to proton charge and electron charge, the proton charge is called "positive" (+), and the electron charge is called "negative" (–), in order to convey the fact that these two particles are equal and opposite in their electrical effects.

Looking again at Fig. 5.10, we see that ten of the electrons are in the two shells close to the nucleus — they are in "low orbits." So, counting just the nucleus and the two inner shells, we have ten electrons and fourteen protons, giving us a net charge of +4. In the *outer* shell, we have four electrons, giving that shell a total (negative) charge of - 4. So, the - 4 of the outer shell balances the +4 charge of the core — the nucleus and inner two shells — leaving the whole atom electrically neutral.

Protons and electrons orbit the nucleus at different levels. Electrons of the outermost orbit interlock with those of neighboring atoms to bind the atoms into a crystal.

Although the silicon atom has *four* electrons in its outer orbit, it has what we can think of as a "desire" to have *eight* electrons in this outer orbit. This desire is what binds silicon atoms together into a crystal. Figure 5.11 is a conceptual diagram of a typical

Figure 5-11.
Silicon Crystal Structure

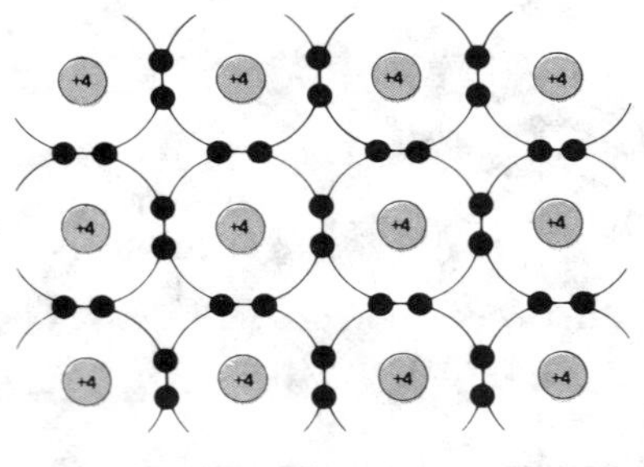

Atoms interlock by sharing electrons in outer orbit.

section of a silicon crystal. Note that each silicon atom's urge to merge has been satisfied, by the clever scheme of sharing each of its outer electrons with four neighbors. So the outer orbit of each atom, in effect, interlocks with the outer orbits of four adjoining atoms.

Since the total number of protons now equals the total number of electrons, the crystal is electrically neutral. Like people whose every desire is satisfied, the atoms just sit in their crystal, smug and content.

From an electrical standpoint, this pure silicon is worthless. It's an insulator — a very poor conductor of electricity — because all of the electrons are tightly *bound* in their shared orbits. These electrons cannot *flow* to carry electrical current.

HOW ARE USEFUL SILICON CRYSTALS MADE FOR SEMICONDUCTORS?

What we need for our diode is not the pure silicon crystal we have seen in Fig. 5.11. We need a crystal with some extra elec-

Pure silicon does not conduct electricity. Silicon prepared for use as a semiconductor material has small amounts of impurities added to create either a surplus or a deficiency of electrons.

trons or extra places for electrons to go. So, while the silicon is still in molten form, we mix in certain impurities intentionally. This is a process called "doping." If we mix, for example, the element phosphorus with the silicon, we get N-type crystalline material. If we use boron, we get P-type. These dopant atoms replace a few of the silicon atoms in the crystal lattice.

Let's take the N-type first. A crystal doped with phosphorus has phosphorus atoms scattered throughout its lattice. They occupy places normally filled by silicon atoms. Now, a phosphorus atom is much like a silicon atom, except that it has one more proton in its core, and a fifth electron in its outer orbit. But where can this lonely fifth electron go? All of the silicon atoms' outer shells are filled with their quota of eight shared electrons. All of the *other* phosphorus atoms have the same problem of an extra electron, so they certainly won't accept another. All the free electrons can do is wander about among the atoms, looking for unfilled orbits. Do not be saddened by their plight, though, for we shall soon occupy their time.

N-type silicon contains phosphorus to create a surplus of free electrons. The material is neutral in charge, but has free negatively charged (hence N-type) electrons to conduct current.

By doping the silicon raw material with phosphorus, we have created a silicon crystal that will conduct electricity. Visualize a generator with a conductor attached to one end of the N-type crystal, and another conductor leading out of the other end of the crystal. We pump electrons into one side of the crystal. Now, recall that like particles repel each other. So for every electron pumped in, one of the free electrons from the phosphorus atoms migrates to the other end of the crystal and out the wire.

What happens when we use a crystal that has been doped with *more* phosphorus? The more phosphorus, the more free electrons we have in the crystal; the more free electrons, the more incoming electrons we can accommodate. The number of electrons *in the crystal* remains constant; one electron leaves for every one pumped in. Multiply what we have shown happens to a single electron, by trillions of times, and you have the concept of electricity flowing in an N-type semiconductor.

HOW IS A P-TYPE SILICON CRYSTAL MADE?

To create P-type silicon crystals, we dope the molten material with *boron* atoms to replace some of the silicon atoms in the crystal lattice. A boron atom has only three electrons in its outer shell. Instead of donating an extra free electron as the phosphorus did, the boron atom creates a deficiency of one electron in this structure. We call this deficiency a "hole."

P-type silicon contains boron to create holes (a deficiency of free electrons). As with the N-type, the P-type material is neutral in charge, but has free voids (holes) which appear as positively charged particles that conduct current.

Just as the free electrons can wander, so can these holes wander about through the lattice, because electrons in the outer shell of the silicon atoms find it very easy to shift over and fill one of these holes. Obviously, a hole is not a physical entity like an electron, but when an electron moves from one place to another, it's just as though the hole it moved to had moved in the opposite direction. A hole always represents a positive charge (+1) after it has moved away from the boron atom. So, the hole can be thought of as a freely moving positive charge.

HOW DOES P-TYPE SILICON CONDUCT ELECTRICITY?

As we did with the N-type silicon crystal, suppose we connect a crystal of P-type silicon to a generator by wires from each side of the crystal. We pump out a bound electron. Now we have a new hole with a positive charge. Recall that like charges repel. So the new hole repels an old hole all the way to the other end of the crystal. At this end, another electron comes in, and the hole gets filled. Thus, bound electrons move one way, by means of holes moving the other way.

As is the case with N-type, when we dope the silicon more heavily with boron atoms, to make it P-type, we increase its current-carrying capacity. This happens because each boron atom creates a hole, and the more holes we have, the more electrons they can accommodate, and the more current the crystal can carry. And again, if we multiply this process trillions of times, we can visualize electricity being conducted by the movement of holes in P-type semiconductors.

Current in N-type material consists of electron flow. Current in P-type material consists of hole flow.

Remember this very important fact: P-type conducts electricity *only* by means of holes, and *has virtually no free electrons.* N-type conducts only by means of free electrons; it has *virtually no holes.*

It's probably obvious now that "N"-type stands for "negative" because of its negatively charged free electrons. "P"-type stands for "positive" because of its positively charged holes.

HOW DO DIODES CONDUCT ELECTRICITY?

Now that we understand how electricity is conducted in N-type and P-type silicon, we are ready to see how these properties are put to use in diodes — to carry electricity in one direction while blocking it in the other.

Figure 5.12 shows how a diode is constructed. Leads are connected to the two ends of a tiny single chip of silicon which is processed to be N-type on one side and P-type on the other. In this example, we have four free electrons in the N material. We have four holes in the P material. The dividing line between the two types is called the "PN junction." It is the behavior of the electrons and holes in the vicinity of this junction that gives diodes and other semiconductors their unique properties.

A diode is made of both types of material. Current flow occurs when an outside voltage source forces electrons and holes toward the "PN junction," causing them to recombine across the junction.

Suppose that, in Fig. 5.12, electrons are being pumped by a generator into the N region. These negatively charged electrons repel the free negative electrons already there, forcing them toward the PN junction. At the same time, bound electrons are being withdrawn from the P region, creating new holes. The new holes repel the old holes, moving the holes toward the PN junction. So the *holes* in the P-type silicon and the *free electrons* in the N-type silicon are moving toward each other. When the holes and free electrons meet at the junction, the free electrons fall into the holes. This conduction process continues as long as new holes and free electrons are being "pumped in."

So that's how a diode *conducts* electricity in one direction. Now, let's see how the device *blocks* the current when it tries to go in the other direction.

Figure 5-12.
Diode Cross-Section—
Good Conduction

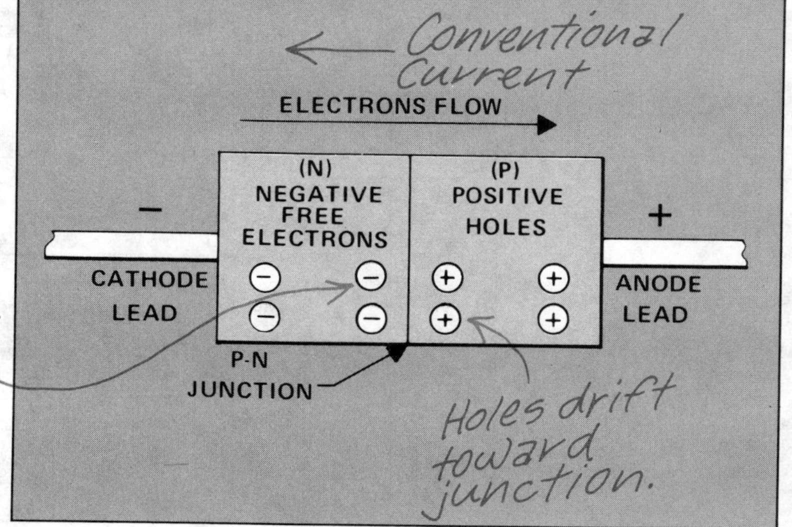

HOW DO DIODES BLOCK ELECTRICITY?

Current flow halts when an outside voltage source pulls electrons and holes away from the PN junction.

Figure 5.13 duplicates Fig. 5.12, except that the electrons are attempting to flow in the opposite direction, from P to N. This is what happens when an ac generator enters the second half of the alternating-current cycle. Since the electrons are attempting to flow from P to N, the free electrons in the N region migrate *away* from the PN junction. In the P region, as bound electrons move in the direction of attempted electron flow, the holes move in the opposite direction, *away* from the junction. The result is there are no free electrons or holes anywhere near the junction. In effect, this zone is like pure, undoped silicon crystal. It is effectively an insulator *until* the electrons again attempt to flow in the acceptable direction. Electrons *can* flow from N to P, but *not* from P to N.

Now you can understand the meaning of the word "semiconductor." This chip of doped silicon is an electrical conductor under certain conditions, but it is an insulator under other conditions. Hence, the term "*semi*conductor."

An understanding of diodes is so important in laying a foundation for our understanding of other semiconductors, that the next chapter is also devoted to diodes. It covers their most significant ratings and characteristics.

Figure 5-13. Diode Cross-Section—Blocked Conduction

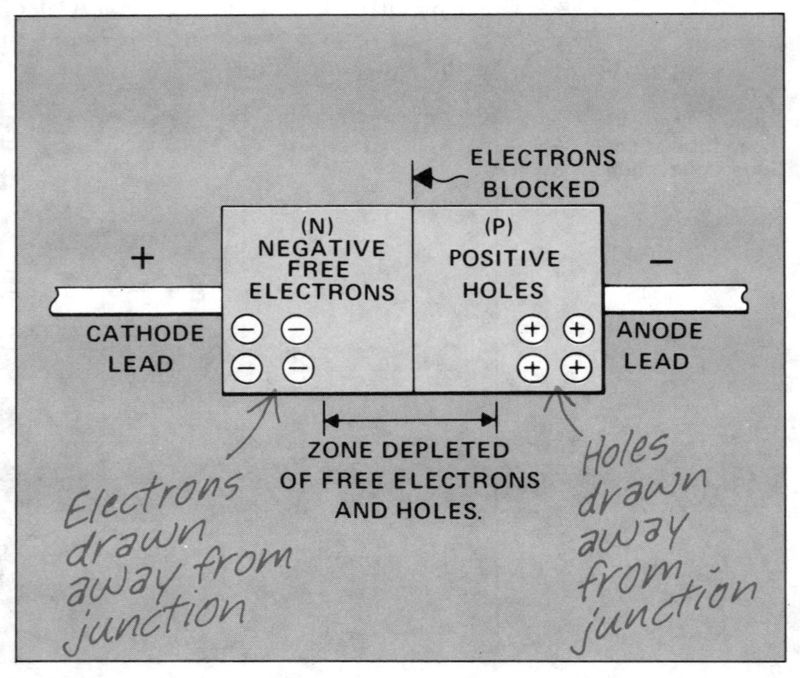

Quiz for Chapter 5

1. Learning the fundamentals of diode operation is very helpful because:

☐ a. Diodes are the simplest type of semiconductor
☐ b. The operation of diodes is applicable to the fundamental operation of all other types of semiconductors
☐ c. Diodes perform both switching and amplifying functions
☐ d. None of the above
☐ e. a and b, above

2. Diodes derive their name from the fact that:

☐ a. They have only two electrodes or electrical connections
☐ b. They can control two different circuits at the same time
☐ c. They are both a regulator and a switch
☐ d. They can operate within two distinct voltage ranges
☐ e. None of the above

3. As a rule, a higher-power, lower-frequency diode is__________ than a low-power, high-frequency diode

☐ a. Larger
☐ b. Smaller
☐ c. Smoother
☐ d. Heavier
☐ e. Cleaner

4. The schematic symbol for a diode is:

☐ a.

☐ b.

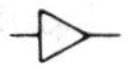

☐ c.

- ☐ d. 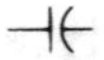
- ☐ e.

5. The basic function of a diode is to:

- ☐ a. Act as a one-way valve for the flow of electrons
- ☐ b. Act essentially as a switching-type device
- ☐ c. Enable electrons to flow in one direction but to be blocked from flowing in the opposite direction
- ☐ d. None of the above
- ☐ e. a, b and c above

6. Power rectification is one function of a diode and in this situation it:

- ☐ a. Converts alternating current to direct current
- ☐ b. Converts direct current to alternating current
- ☐ c. Can be coupled with a capacitor to provide a relatively smooth source of direct current
- ☐ d. None of the above
- ☐ e. a and c, above

7. When a diode is used in a circuit to prevent the voltage in one wire from exceeding the voltage in a second wire it is being used as a:

- ☐ a. Clamping diode
- ☐ b. Part of a logic gate
- ☐ c. Detector diode
- ☐ d. Rectifier
- ☐ e. All of the above

8. A diode used as a detector (demodulator) in a crystal radio receiver:

- ☐ a. Detects the presence of electrons in the air
- ☐ b. Detects metal in the earth
- ☐ c. Detects direct current
- ☐ d. Rectifies high-frequency amplitude-modulated (AM) alternating current from the antenna, producing pulses of direct current which "average out" as low-frequency sound signals
- ☐ e. Detects mistakes in circuit design

9. Even though N-type semiconductor material has some freely moving electrons in addition to the bound ones, and P-type material has empty spots ("holes") among its bound electrons, still the material is electrically neutral, with no excess or deficiency of electrons, because;

- ☐ a. Every atom that went to make up the material had an equal number of negative electrons and positive protons, and no electrons or protons have been gained or lost
- ☐ b. Every silicon atom has four more protons than electrons, and this inequality is balanced out by doping
- ☐ c. All the free electrons in N-type material are pumped out by the electric circuit in which the material is used
- ☐ d. All the holes in P-type material are filled up by electrons pumped in by the electric circuit in which the material is used
- ☐ e. Phosphorus atoms have one more electron than protons, and boron atoms have one less electron than protons, and equal quantities of these two dopants are always used in N-type or in P-type material

10. As electrons are pumped through P-type material,

- ☐ a. Holes move in the opposite direction
- ☐ b. Bound electrons flow — there are virtually no free electrons in P-material
- ☐ c. A bound electron can only move when there is a hole next to it for it to move into
- ☐ d. The net effect is as if some sort of positive particles were moving in the opposite direction
- ☐ e. All of the above

11. A PN junction acts as a one-way valve for electrons because:

- ☐ a. When electrons are pumped from N to P, free electrons and holes are forced together, so free electrons fall into holes near the junction
- ☐ b. When we attempt to pump electrons from P to N, free electrons and holes are forced apart, leaving no way for electrons to cross the junction
- ☐ c. There is a little mechanical switch inside each diode
- ☐ d. The circuit in which the diode is used only attempts to pump electrons in one direction
- ☐ e. a and b above

Key Words

Conventional Current
Forward Current (I_F)
Forward Voltage (V_F)
Reverse Breakdown Voltage $V_{(BR)}$
Reverse Current (I_R)
Reverse Recovery Time (t_{rr})
Reverse Voltage (V_R)

Definitions are found in the glossary in the back of the book.

Diode Performance and Specifications

Before resuming our discussion of diodes, we have to take a side trip and return to basic electricity, to pick up a concept that was introduced but needs further explanation.

ARROWS IN SEMICONDUCTOR SYMBOLS POINT IN THE DIRECTION OF CONVENTIONAL CURRENT

Figure 6.1 shows symbols for a diode and an NPN transistor. We've seen them both before. The diode allows the electrons to

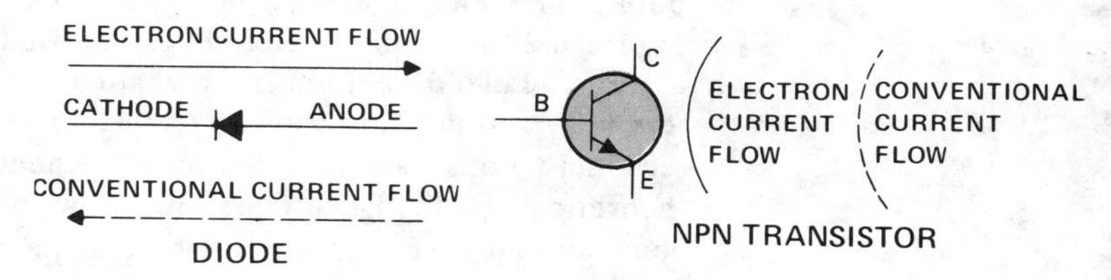

Figure 6-1. Current Flow Through a Diode and a Transistor

The arrows in semiconductor symbols point in the positive-to-negative direction of conventional current flow (opposite the direction of electron flow).

flow in a direction opposite to the way the arrow points; electrons flow from cathode to anode. Similarly, the transistor allows electrons to flow from emitter to collector when electrons are withdrawn from the base, but the arrow points the other way.

As discussed previously, the arrow does point in a significant direction, however. It points in the direction of *conventional* current flow.

WHAT IS CONVENTIONAL CURRENT?

Benjamin Franklin was an imaginative scientist, but he once made a wrong guess and invented "conventional current." In his time, little more was known about electricity than that, if we rubbed amber and glass with certain substances such as wool and silk, the amber and glass attracted each other by means of some mysterious physical force. On the other hand, two pieces of rubbed glass repelled each other, and two pieces of rubbed amber repelled each other. (Unlike charges attract, like charges repel).

Franklin's explanation was satisfactory for that time. He said that all bodies, including the glass and amber, contain a mysterious invisible fluid. This fluid became known as electricity. The theory held that amber and glass each have a natural amount of this fluid, which remains constant under ordinary conditions. But rubbing the amber and glass takes fluid away from one substance and gives fluid to the other substance. The attraction of the unlike bodies was explained as the tendency of the fluid to return to its normal level in each body. The excess fluid was termed a positive charge, and conversely, a deficiency of fluid was called a negative charge. Franklin's theory said that the electrical fluid would tend to flow through a wire from a region of positive charge to a region of negative charge. But he didn't know *which* body — the amber or the glass — had the excess, and which had the deficiency, of fluid.

Conventional current was an invention of Benjamin Franklin, who made hypotheses about the nature of electricity based on his experiments with glass and amber.

So he made an intuitive guess. He guessed that the glass had the excess, and was therefore the positively charged body, and he labeled the charge on the amber as negative. His guess became the *convention* in all of the electrical theory, mathematics, textbooks, and electrical equipment for the next hundred years. Figure 6.2 shows how a battery supposedly pumped Franklin's theoretical electrical fluid from what came to be called its positive terminal, through the motor, to what was labeled the negative terminal. Volts became the measurement of pressure of this theoretical fluid. Figure 6.3 shows how, according to Franklin's conventional theory, a conductor can have any voltage ranging from low negative voltages (deficiency of fluid) to high positive voltages (excess of this fluid).

Ben was right in saying that electricity is a sort of fluid, but wrong in his guess as to which way it flows. But it was a century before anybody determined this. When they did, the scientists discovered that the fluid really flowed from what was con-

Figure 6-2. Direction of Flow for Conventional Current and Electron Current

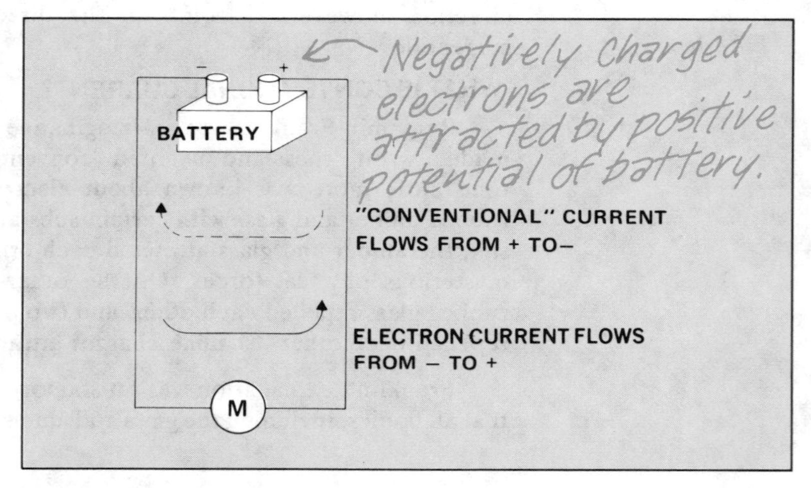

Figure 6-3. Conventional Current and Electron Current Flow Under Influence of Applied Voltage

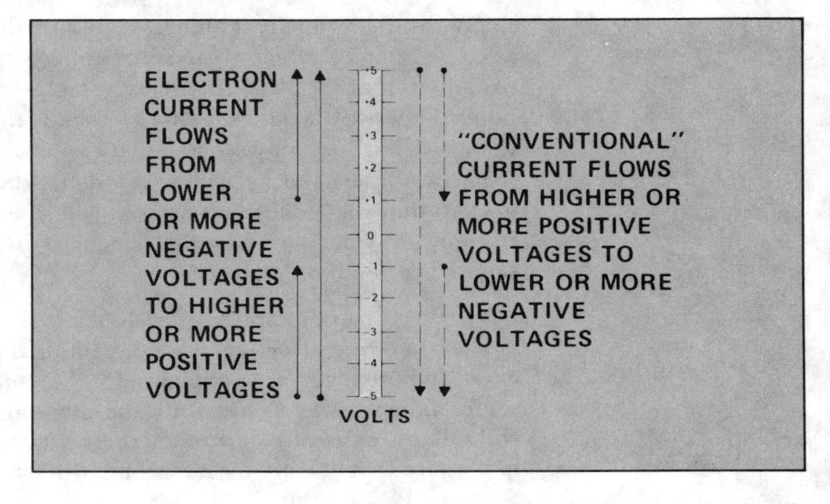

ventionally called "negative," to the conventional "positive." Fig. 6.2 and 6.3 show how the *electron* flow is really opposite to that of Franklin's current.

To make the electrons match with Franklin's theory, which was now accepted worldwide, the electrons had to be called negative. So all electrical motors and equipment, textbooks, and mathematics showing electricity flowing from *positive to negative* were in reality showing the flow of an *imaginary* fluid. This imaginary fluid is what we now call "conventional" current.

Although later investigations showed that electron current flows from negative to positive, it was expedient to preserve the convention of positive-to-negative current flow in circuit analysis.

It's easy to say that the scientists should simply have redefined Franklin's assignment of positive and negative, to make the electrons positive. But in those days, it would have been just as confusing and expensive as converting the U.S. from the English to the metric system today. Moreover, the idea of an imaginary kind of electricity flowing from positive to negative is perfectly satisfactory. Positive current in one direction is exactly equivalent to negative current in the other direction. A good example is our earlier discussion of holes in semiconductor material. Holes have the effect of positive charges moving one way — but nothing positive is moving. Indeed, nothing *physical* is moving in that direction; all that is moving is the *location* of holes, as electrons move in the opposite direction.

Even today, most engineers think and talk in terms of Franklin's conventional current. Generally, when an engineer mentions current or voltage, he means *conventional* current or voltage. When he is referring to *electron* current, he will normally specify it as electron current. When he talks about electron *voltage* he will usually refer to it as *negative* voltage.

We didn't want to confuse you with this, but if you'll think about it, it will really clear some things up. Think back to our explanation of the NPN transistor. An electrical engineer (and the books he writes) would say that current is flowing *into* the base of the transistor—when we know that *electrons* are being *withdrawn* from the base. And he would say that, when the NPN transistor goes to the "on" condition, current will flow from the collector to the emitter, when we know that electrons in fact are moving in the *opposite* direction.

In this chapter, and when we discuss the applications of other devices in circuits, we will use conventional current terms. But in cases where we are explaining the *internal* operation of semiconductors, as we did with the diode in the last chapter, we will talk in terms of *electron* current. In each case, though, we'll make sure that you are aware of our switch in terminology.

So prepared, we are ready to return to our discussion of diodes, and discuss diode behavior.

WHAT IS MEANT BY DIODE BEHAVIOR?

Diode behavior can be described in terms of the voltage across the diode and the current through the diode.

A diode will pass current in the forward direction and block current in the reverse direction as shown in Fig. 6.4. Notice that the direction is now in terms of conventional current. The diode's behavior is the relationship between voltage (V) and current (I). Forward voltage (V_F) is defined as the amount by which the anode voltage exceeds the cathode voltage. (Remember we are speaking of positive voltage — the pressure of Franklin's imaginary fluid.) In Fig. 6.4, the forward voltage is 1 volt, since the anode voltage (+6) exceeds the cathode voltage (+5) by 1 volt. Forward current (I_F) is simply the amount of current at a given forward voltage.

Since no diode is perfect, current is not always completely blocked in the reverse direction. When a reverse voltage (V_R) is

Figure 6-4. Diode Behavior with Conventional Current

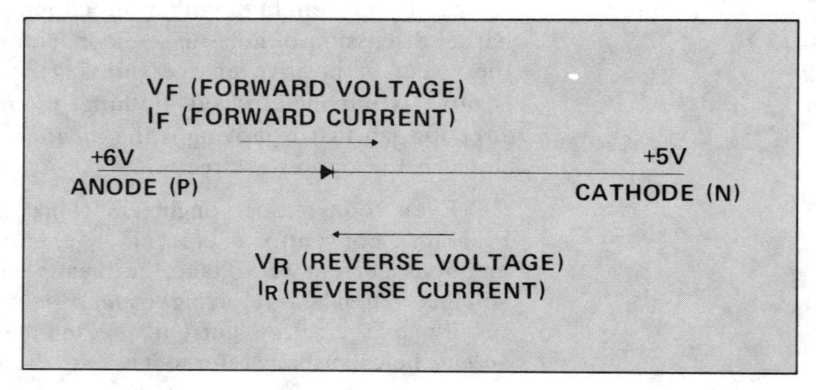

applied, a small amount of reverse current (I_R) will flow. Reverse voltage is the amount by which the *cathode* voltage exceeds the *anode* voltage. And reverse current is the amount of current at a given reverse voltage.

HOW CAN DIODE BEHAVIOR BE SHOWN GRAPHICALLY?

The curve in Fig. 6.5 is the key to understanding diode specifications. Nearly all of the important diode characteristics can be determined from this behavior curve of a typical diode.

Figure 6-5. Characteristic Curve for a Diode

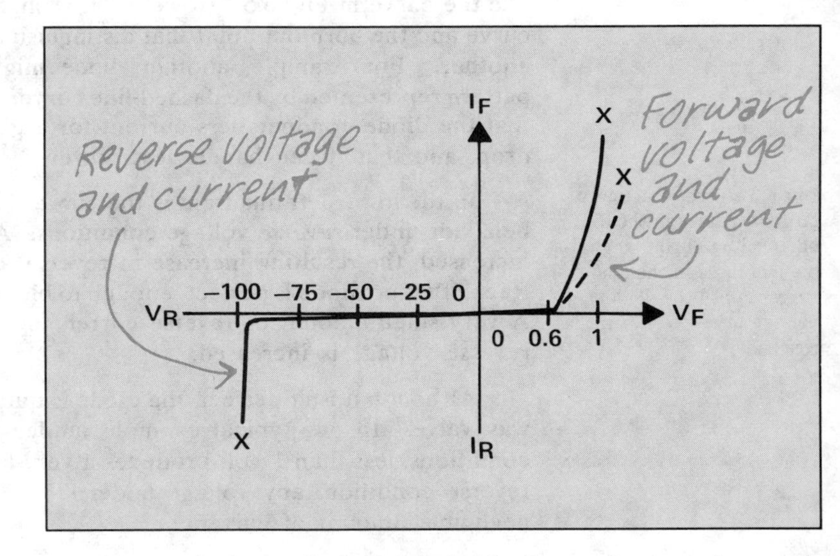

Characteristic curves of the diode's voltage plotted against the diode's current graphically represent a diode's performance under various combinations of voltage and current.

The vertical coordinate is labeled I_F (forward current) toward the top, and I_R (reverse current) toward the bottom. The horizontal coordinate is labeled V_F (forward voltage) toward the right, and V_R (reverse voltage) toward the left. Notice that the scale relationship changes considerably between the forward voltage and reverse voltage sides; one volt of V_F on this particular graph is equal in distance along the axis to -50 volts of V_R.

Silicon diodes have a forward threshold voltage of 0.6 volt. Germanium diodes have a threshold voltage of 0.3 volt.

The axes divide the area into four quadrants. The upper right-hand quadrant shows the behavior of the diode under forward-voltage conditions. At zero forward voltage, forward current is zero. As the forward voltage is increased, the current increases very gradually at first, and then more rapidly. This curve has a fairly pronounced "knee" in the vicinity of a certain voltage (0.6 volt in silicon diodes). At this forward voltage, the *current* begins to increase dramatically, and the curve turns upward. This point can be thought of as the threshold voltage, where the diode really begins to turn on. (In germanium diodes, the point is at about 0.3

Beyond the threshold voltage in the forward direction, small increases in the diode's voltage cause large increases in the diode's current.

volt.) From this point on, slight increases in voltage cause greater and greater increases in current. The upward curve ends abruptly at a limit where the diode burns out due to power dissipation — heat generated in the diode. Recall that power dissipation (measured in watts) is equal to current (in amps) multiplied by voltage drop (in volts). So as current and voltage are increased, more and more heat is generated until, at some point, the diode burns up. Power dissipation is abbreviated "P."

Every diode has a forward conduction curve that looks much like the curve in Fig. 6.5. However, it is slight differences in this curve and the burn-out point that distinguish one diode type from another. For example, another diode might have a behavior pattern represented by the dashed-line curve. This curve indicates that the diode conducts less current for a given forward voltage drop, and that it can handle less power.

In the reverse direction, large increases in the diode's voltage bring negligible increases in the diode's current until the diode's voltage breakdown point is reached.

In the lower left quadrant is the curve that shows the diode's behavior under *reverse* voltage conditions. As reverse voltage is increased, the resulting increase in reverse current is as shown. Recall that no diode is perfect enough to block *all* reverse current. A very small amount of reverse current does "leak through" as reverse voltage is increased.

Although it isn't perfect, the diode is quite effective as a one-way valve. In our typical example, under forward conduction conditions, less than 1 volt produces a very large current. But in reverse condition, any voltage under -75 volts produces only a negligible amount of current.

In the reverse condition, a point is finally reached where the blocking capability starts to break down completely. Beyond this point, called $V_{(BR)}$ for "reverse breakdown voltage," the diode cannot hold back the reverse current. At this point, current increases dramatically and the diode soon burns up. This happens quickly because, at these relatively high voltages, it takes very little current to generate a high wattage of destructive heat.

WHAT DIODE SPECIFICATIONS ARE MOST IMPORTANT?

You're probably ahead of us, but let's consider what some of the more important diode specifications are. Although there are more than a dozen ways to specify diodes, just five specifications are really important:

I_F *(forward current)*, the amount of current the diode can handle without burning up; a measure of how much power the device can dissipate.

V_F *(forward voltage)*, the voltage level necessary to produce the desired forward current level.

I_R *(reverse current)*, the amount of current that will leak through the diode at various reverse voltages.

$V_{(BR)}$ *(reverse breakdown voltage)*, the reverse voltage beyond which current begins to rise very rapidly.

t_{rr} *(reverse recovery time)*, the time it takes the diode to recover from forward conduction and begin to block reverse current. This becomes important when we consider frequency; the higher the frequency of alternating current being imposed on the diode, the quicker the diode must respond in order to rectify this current.

Four of the five very important diode specifications can be read directly from the diode's characteristic curve.

All of these specifications (except t_{rr}) can be read directly from an I-V curve, a current-voltage curve of the sort we have just seen. Figure 6.6 shows where each of these points is located on a typical curve.

It's important to note that these curves are derived by testing devices under certain fixed *temperature* conditions, because the curve will change as temperature is changed. Typically, the device

Figure 6-6. Characteristic Curve Showing Diode Parameters

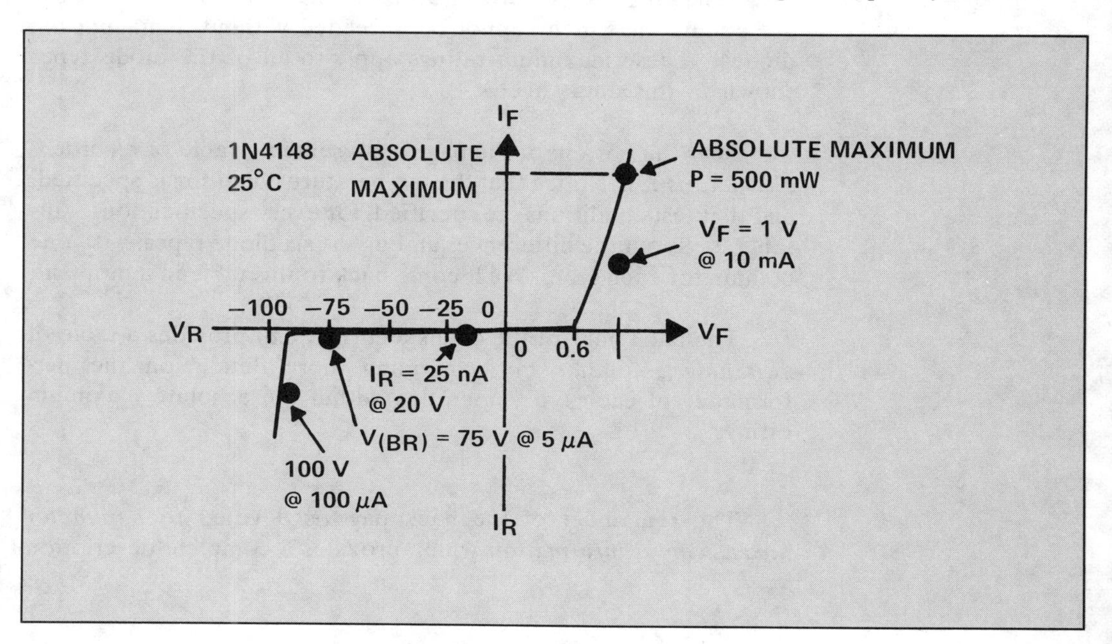

Manufacturers obtain the characteristic curves of semiconductor devices at controlled temperatures. Changes in temperature cause current through the devices to vary somewhat.

is surrounded by air at 25°C, which is normal room temperature. If, for example, the air is instead at 100°C, we would get considerably more current for a given voltage drop. The increased temperature would cause an increase in forward current, and an increase in reverse leakage current.

HOW ARE THESE SPECIFICATIONS SHOWN ON A DATA SHEET?

Data sheets summarize diode characteristics with graphs, drawings, lists of specifications, and tables of parameter values.

We're ready now to tackle a data sheet. Data sheets are really much simpler than they look. Figures 6.7 and 6.8 are the front and back of a complete diode data sheet. All diode manufacturers present the data in much the same way — we just happened to select an old data sheet from Texas Instruments Incorporated because we had one handy. TI no longer manufactures this device.

Much of the standard data sheet is self-explanatory. The heading shows that this one covered a whole family of planar silicon switching diodes; the 1N4148 whose curve we have shown, plus five other very similar diodes. Below this is presented a summary of the most prominent characteristics of the diodes. Below that are the mechanical data, a description of the construction of the diode and an engineering drawing of the package.

Following this is a list of *absolute maximum ratings.* These are conditions which cannot be exceeded without damaging the diodes. These maximum ratings apply to all of the diode types shown in this data sheet.

Near the bottom of the page is presented a table of *electrical characteristics.* Notice that the temperature condition is specified, and that test conditions are specified, for every specification. This table spells out the differences among the six diode types, with one column for each type. We'll come back to this table in a moment.

The back page of the data sheet (Fig. 6.8) provides a table of *switching characteristics,* providing more details on the performance of each type, operating within the absolute maximum ratings.

The remainder of the back page is devoted to *parameter measurement information* which provides a complete description of exactly how the tests were made on each diode — so that the purchaser may verify the specifications for himself.

HOW CAN WE RELATE THESE DATA SHEET SPECIFICATIONS TO THE I-V CURVE?

The information appearing on the data sheet is taken from various parts of the diode's characteristic curve.

Refer back to Fig. 6.6 as you read the data sheet, and you'll see which parts of the curve the data are taken from.

Power dissipation (P) can be found in the *absolute maximum ratings* table. The rating is shown as 500 mW (milliwatts). This point is shown near the top of the curve of Fig. 6.6. Sometimes, this limit is specified instead in terms of *absolute maximum forward current.*

Forward voltage (V_F) is found in the table of *electrical characteristics.* It shows that static forward voltage for the 1N4148 was tested with forward current at 10 mA (milliamps). At a forward current of 10 milliamps, the 1N4148 has a forward voltage not exceeding 1 volt. This specification is also shown on the curve, and in effect, says TI guarantees that the curve for every 1N4148 will pass to the left of this point.

Reverse current (I_R)is also found in the *electrical characteristics* table. There are two test conditions: First, at a reverse voltage of 20 volts and air temperature of 25°C, there is a reverse current not exceeding 25 nA (nanoamps, or billionths of an amp). This also appears as a point on the curve. In effect, it guarantees that the curve for every 1N4148 will pass above this point. The second test condition for I_R is at 20 volts and 150°C. Here, we see how reverse leakage current increases with temperature — at 150°C, the maximum leakage increases to 50 μA (microamps, or millionths of an amp). You won't find this point on the curve, of course, since the curve covers only the 25°C test conditions.

Reverse breakdown voltage ($V_{(BR)}$) is in the same table, and two different limit points are specified. First, with a reverse breakdown current of 5 μA, reverse voltage will be no less than 75 volts. Second, the specifications says that when breakdown current reaches 100 μA, reverse voltage will be no less than 100 volts.

Reverse recovery time (t_{rr}) is found in the table of *switching characteristics.* The test conditions shown, incidentally, are fairly

Figure 6-7.
Diode Data Sheet (Front)

TYPES 1N4148, 1N4149, 1N4446 THRU 1N4449
SILICON SWITCHING DIODES

BULLETIN NO. DL-S 739269, OCTOBER 1966–REVISED MARCH 1973

FAST SWITCHING DIODES

- **Rugged Double-Plug Construction**

Electrical Equivalents:

1N4148 . . . 1N914 . . . 1N4531	1N4447 . . . 1N916A
1N4149 . . . 1N916	1N4448 . . . 1N914B
1N4446 . . . 1N914A	1N4449 . . . 1N916B

mechanical data

Double-plug construction affords integral positive contact by means of a thermal compression bond. Moisture-free stability is ensured through hermetic sealing. The coefficients of thermal expansion of the glass case and the dumet plugs are closely matched to allow extreme temperature excursions. Hot-solder-dipped leads are standard.

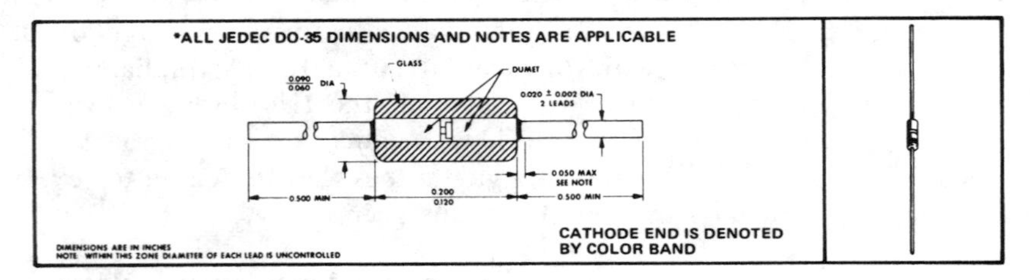

***absolute maximum ratings at 25°C free-air temperature (unless otherwise noted)**

$V_{RM(wkg)}$	Working Peak Reverse Voltage	75 V
P	Continuous Power Dissipation at (or below) 25°C Free-Air Temperature (See Note 1)	500 mW
T_{stg}	Storage Temperature Range	–65°C to 200°C
T_L	Lead Temperature 1/16 Inch from Case for 10 Seconds	300°C

***electrical characteristics at 25°C free-air temperature (unless otherwise noted)**

PARAMETER	TEST CONDITIONS	1N4148		1N4149		1N4446		1N4447		1N4448		1N4449		UNIT
		MIN	MAX	MIN	MAX	MIN	MAX	MIN	MAX	MIN	MAX	MIN	MAX	
$V_{(BR)}$ Reverse Breakdown Voltage	$I_R = 5\ \mu A$	75		75		75		75		75		75		V
	$I_R = 100\ \mu A$	100		100		100		100		100		100		V
I_R Static Reverse Current	$V_R = 20$ V		25		25		25		25		25		25	nA
	$V_R = 20$ V, $T_A = 100°C$										3		3	µA
	$V_R = 20$ V, $T_A = 150°C$		50		50		50		50		50		50	µA
V_F Static Forward Voltage	$I_F = 5$ mA									0.62	0.72	0.63	0.73	V
	$I_F = 10$ mA		1		1									V
	$I_F = 20$ mA						1		1					V
	$I_F = 30$ mA												1	V
	$I_F = 100$ mA										1			V
C_T Total Capacitance	$V_R = 0$, $f = 1$ MHz		4		2		4		2		4		2	pF

NOTE 1: Derate linearly to 200°C at the rate of 2.85 mW/°C.

* JEDEC registered data

Figure 6-8.
Diode Data Sheet (Back)

TYPES 1N4148, 1N4149, 1N4446 THRU 1N4449
SILICON SWITCHING DIODES

***switching characteristics at 25°C free-air temperature**

PARAMETER		TEST CONDITIONS	1N4148		1N4149		1N4446		1N4447		1N4448		1N4449		UNIT
			MIN	MAX	MIN	MAX	MIN	MAX	MIN	MAX	MIN	MAX	MIN	MAX	
t_{rr}	Reverse Recovery Time	I_F = 10 mA, V_R = 6 V, i_{rr} = 1 mA, R_L = 100 Ω, See Figure 1		4		4		4		4		4		4	ns
$V_{FM(rec)}$	Forward Recovery Voltage	I_F = 50 mA, R_L = 50 Ω, See Figure 2										2.5		2.5	V

*PARAMETER MEASUREMENT INFORMATION

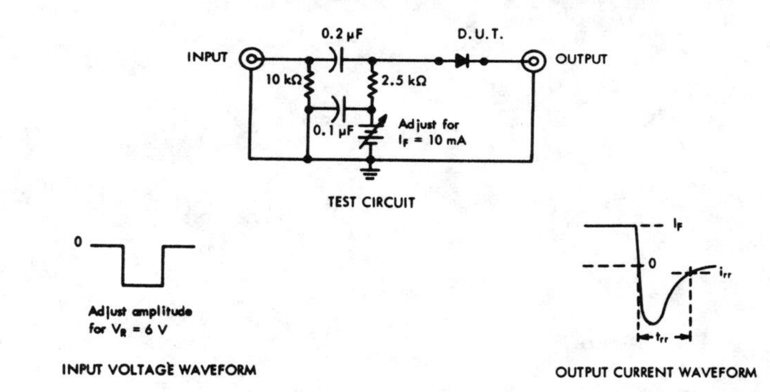

FIGURE 1 — REVERSE RECOVERY TIME

NOTES: a. The input pulse is supplied by a generator with the following characteristics: Z_{out} = 50 Ω, $t_r \leq$ 0.5 ns, t_p = 100 ns.
b. The output waveform is monitored on an oscilloscope with the following characteristics: $t_r \leq$ 0.6 ns, Z_{in} = 50 Ω.

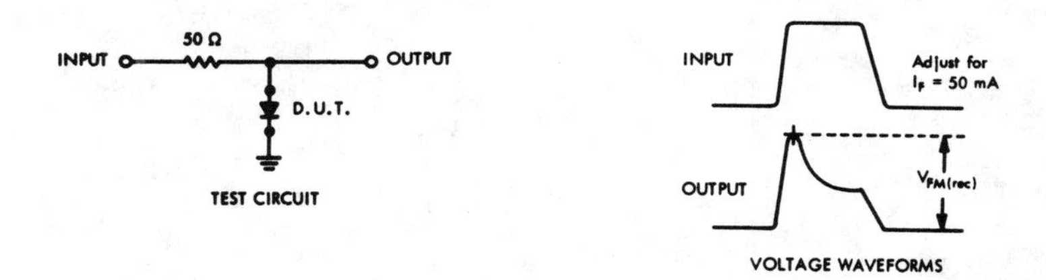

FIGURE 2 — FORWARD RECOVERY VOLTAGE

NOTES: c. The input pulse is supplied by a generator with the following characteristics: Z_{out} = 50 Ω, $t_r \leq$ 30 ns, t_p = 100 ns, PRR = 5 to 100 kHz.
d. The output waveform is monitored on an oscilloscope with the following characteristics: $t_r \leq$ 15 ns, $R_{in} \geq$ 1 MΩ, $C_{in} \leq$ 5 pF.

*JEDEC registered data

typical operating conditions. The t_{rr} is shown as 4 ns (nanoseconds, billionths of a second). This is the time that elapses after forward conduction has stopped, before the diode can begin blocking the reverse current.

Armed with these 5 significant diode specifications, you're in a good position to interpret diode data sheet information intelligently — in fact, you know enough about diodes to select them for use in simple circuits. More important, your knowledge of diodes provides an excellent background for a study of transistors, which we begin in the next chapter.

Quiz for Chapter 6

1. A diode is:

☐ a. A one-way valve for the flow of electricity
☐ b. A basic semiconductor used extensively in digital computers
☐ c. A relatively simple and therefore inexpensive semiconductor
☐ d. All of the above
☐ e. a and b above

2. Electron current flows through a diode in this direction:

☐ a.

☐ b.

☐ c.

☐ d.

3. Conventional current flows:

☐ a. From positive to negative
☐ b. From negative to positive
☐ c. In the direction opposite to electron flow
☐ d. a and c above
☐ e. None of the above

4. In a circuit powered by a battery, electrons are being pumped:

☐ a. From negative terminal around to positive terminal
☐ b. From positive terminal around to negative terminal
☐ c. Back and forth from terminal to terminal
☐ d. None of the above

5. Although imaginary, the concept of conventional or positive current is commonly used today because:

- ☐ a. Changing the system would be too troublesome and expensive
- ☐ b. The electrical effect is exactly the same, whether you think of conventional current going one way, or electron current going the other way
- ☐ c. Electrical engineers are generally very comfortable with the concept of conventional current
- ☐ d. All of the above
- ☐ e. None of the above

6. A diode:

- ☐ a. Is a perfect one-way valve and stops the flow of all electrons in the reverse direction
- ☐ b. Does not stop all electrons from leaking in the reverse direction
- ☐ c. A device through which no current flows when voltage on both the cathode and anode are equal
- ☐ d. None of the above
- ☐ e. b and c above

7. Data sheets:

- ☐ a. Contain the key electrical information we need to know about a diode
- ☐ b. Indicate how much logic data a diode can handle
- ☐ c. Contain the key mechanical specifications of a diode
- ☐ d. All of the above
- ☐ e. a and c above

Use the reproduction of a TI data sheet for a 1N4148 diode (Figs. 6.7 and 6.8) to determine certain key performance characteristics at 25°C (room temperature) for questions 8, 9, 10 and 11:

8. The power dissipation rating of this device is:

- ☐ a. 500 mW (milliwatts)
- ☐ b. 75 V (volts)
- ☐ c. 4 V
- ☐ d. 50 μA (microamperes)
- ☐ e. None of the above

9. The forward voltage drop at 10 mA (milliamperes) of current is no more than:

- ☐ a. 1 V
- ☐ b. 75 V
- ☐ c. 2 μ A
- ☐ d. 5 V
- ☐ e. None of the above

10. The reverse breakdown voltage at 100 μA of current is no less than:

- ☐ a. 100 V
- ☐ b. 75 V
- ☐ c. 1V
- ☐ d. 50 μA
- ☐ e. None of the above

11. The reverse recovery time is no more than:

- ☐ a. 1 nanosecond
- ☐ b. 4 nanoseconds
- ☐ c. 1 mA
- ☐ d. Indetermiate
- ☐ e. None of the above.

7 TRANSISTORS: HOW THEY WORK AND HOW THEY ARE MADE

Key Words

Alloy
Diffused Junction
Epitaxial
Mesa Diffusion
Monocrystalline
Planar Diffusion
Polycrystalline

Definitions are found in the glossary in the back of the book.

Transistors: How They Work and How They Are Made

Since it may be a while since you've read Chapter 2, let's take a moment to recall what a transistor does in a circuit.

WHAT DOES A TRANSISTOR DO?

You'll recall that a transistor can be used as an amplifier or a switch. Figure 7.1 illustrates a typical amplifying application — an

Figure 7-1.
Transistor Used as an Amplifier

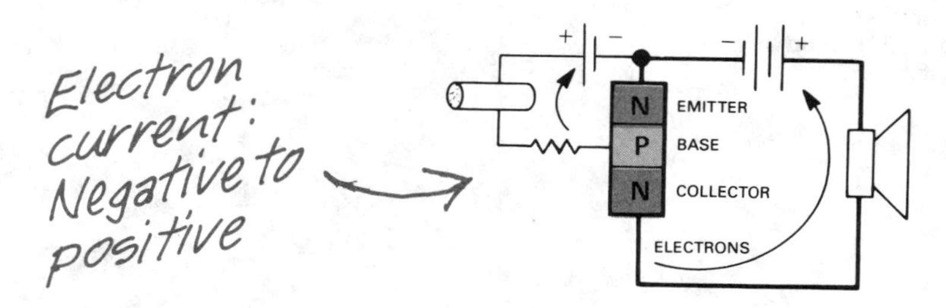

Used as an amplifier, a transistor regulates current flow to track and amplify the fluctuations of a varying input signal.

NPN transistor amplifying current from a microphone to drive a loudspeaker. Speaking now in terms of *electron* current, the microphone converts the power of sound waves into waves of electrical current. The microphone pumps surges of electrons out of the base (P-region) of the transistor. When there is no sound wave activating the microphone, the transistor simply blocks the current from the power supply in the working circuit. But when electrons are pulled out of the P-region by the microphone, a much larger but proportional quantity of electrons will flow from emitter to collector and on through the circuit to the loudspeaker.

You'll recall that we needed the transistor because the current produced by the microphone is so small that it will not drive the loudspeaker directly. So we need more power — and we get it from our power supply, which in this case is a battery. The transistor *regulates* the flow of electrons from the battery, and produces an amplified copy of the much weaker microphone signal.

Based on our discussion of diodes, you can see that this amplification process is due to the movement of free electrons and positive holes in the semiconductor material of the transistor. We'll see how this works in just a moment. But let's recall how a transistor works in its other kind of application, *switching*.

Used as a switch, a transistor alternates between a fully-on and fully-off condition.

Our now-familiar switching transistor application, a telegraph sending and receiving system, using conventional current, is shown in Fig. 7.2. You'll recall that the switch in the control circuit produces dots and dashes, but that the great length of wire between the switch and the receiver has so much resistance that the control-circuit signal is too weak, by the time it arrives, to activate the buzzer. As the control circuit voltage on the base is raised above the switching threshold, the transistor "turns on" a large quantity of current from the working circuit power supply, to activate the buzzer.

Figure 7-2. Transistor Used as a Switch

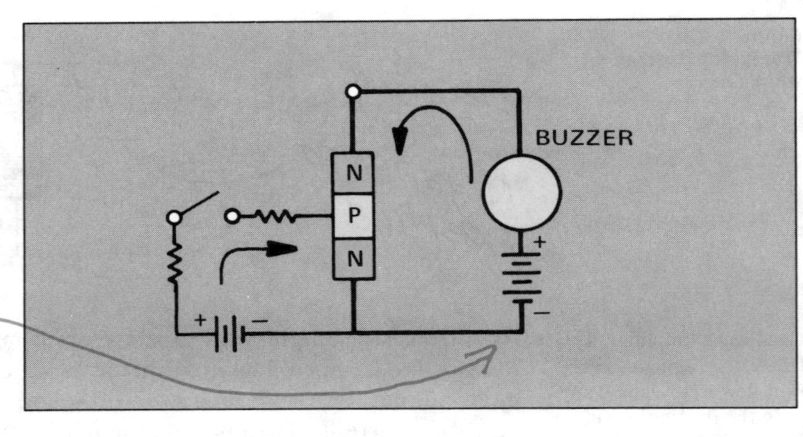

Conventional current: Positive to negative

What is this "threshold" we just mentioned? The threshold voltage of a silicon transistor is about the same as that of a silicon diode — which we said was about 0.6 volt. When we raise the control voltage to the P-region to produce a forward voltage difference of about 0.6 volt across the base-emitter junction, the transistor will go to an "on" condition . . . and cause current to flow in the base-emitter circuit so that a proportionally greater current can flow in the working circuit. More about threshold voltage later.

WHAT IS THE DIFFERENCE BETWEEN A SWITCHING TRANSISTOR AND AN AMPLIFYING TRANSISTOR?

Some transistors are manufactured so they operate better as switches, and others are made so they operate better as amplifiers. But in a pinch, most transistors can be used to either switch or amplify. The transistor in our telegraph system, for example, could function as an amplifier. We would sacrifice good performance in most cases, however. For this reason, transistor types are generally classified as amplifiers or switches, but not both.

The design of the control circuit determines whether a transistor acts as a switch or an amplifier.

It is not the transistor itself that determines whether it will switch or amplify; rather, it is the control circuit, the device that controls the transistor, that causes it to function as one or the other.

HOW DOES A TRANSISTOR WORK?

We promised you that many of the principles of diodes and their behavior would help us in studying other semiconductors. That background now pays off in understanding how a transistor works. Figure 7.3 is a schematic cross-section of an NPN transistor element. You'll remember that N-type semiconductor material conducts electricity by means of its supply of free electrons, and that P-type conducts by its supply of positively charged holes.

Transistor operation centers on the interaction between two PN junctions.

In this discussion, we are talking in terms of electron current. You'll note that the P-region of our transistor is much narrower than the N-regions. This P-region is also much less heavily doped than the N-regions; that is, the holes are fewer and farther apart, compared to the free electrons in the N-regions.

Suppose we pump free electrons from the emitter lead to the collector lead. Assume the base lead is open so that electrons can't flow through the base lead. If you're thinking in terms of our previous discussions on diodes, a surprising thing happens. The free electrons continue merrily on their way, from the N-region of the emitter, across the P-region of the base, into the N-region of the collector, and on down the wire. This process will continue for

Figure 7-3. NPN Transistor Cross-Section

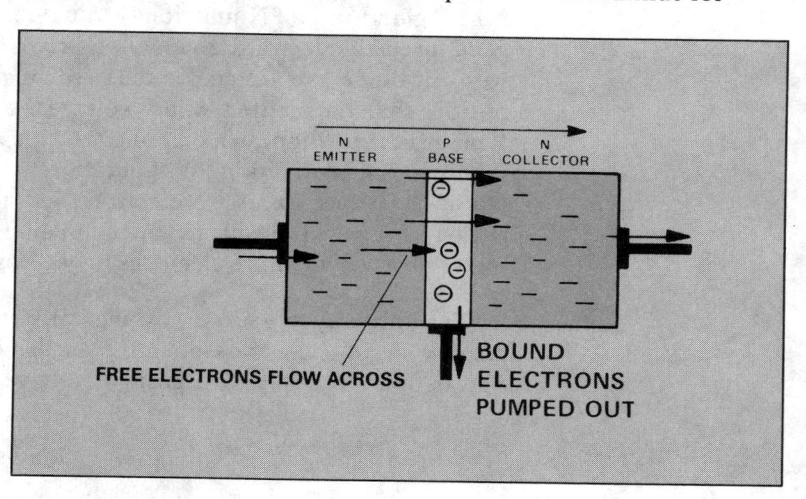

only a brief instant of time, but we need to understand it so we can fully appreciate how the transistor works.

You may be thinking that something is wrong here. The rule we learned in talking diodes is that electrons *can* flow from *N to P*, but they *cannot* flow from P to N. That remains true — so once the electrons get into the base, how can they pass on to the collector?

Electron flow in an NPN transistor passes from emitter to collector via the base. The small accumulation of electrons in the base serves to regulate current through the transistor.

You might expect that these free electrons would be captured in the base area by falling into holes, so that no electrons would pass from the base to the collector. The secret is that the transistor is made with a very *narrow* base region which is very *lightly doped*, so that the holes are scattered sparsely. So most of the emitted electrons — typically, 98% — are able to cross this no-man's-land without falling into a hole.

However, the few electrons that *do* fall into holes are stuck there. They accumulate in the base region, piling up a negative charge in the base. This is what permits the transistor to perform its job of throttling the emitter-collector working current. Remember that like charges repel.

The excess bound electrons in the base region repel the free electrons trying to cross through from emitter to collector, making it harder for this current to pass. It doesn't take very long — perhaps 50 nanoseconds — for current to be shut off entirely.

Let's pause a moment and consider the nature of this barrier that is shutting off current flow. Notice that the emitter and the base region form a PN junction — a diode. We know now that to get appreciable forward conduction across this diode junction, as with any diode, the voltage on the base must be at least 0.6V more positive than the emitter. At this voltage the bound electrons in the P material are withdrawn creating the holes that move toward the junction for conduction. Without this voltage to withdraw the bound electrons, excessive electrons will accumulate in the base region and repel electrons being pumped into the N emitter material. Thus, no more electrons flow across the junction.

The only way to get the working current going again is to withdraw some of the excess bound electrons from the base region. This is done by applying a positive voltage to the base lead, and simply "sucking" out bound electrons into the base-emitter control circuit. This creates new holes in the base, tending to restore the proper number of holes in the P material and reducing the number of repelling electrons. With the repelling barrier lowered, electron current resumes from emitter to collector. For every electron withdrawn from the base, typically 50 electrons cross over from emitter to collector before one falls into a hole. Thus, the small base current proportionately controls the larger working current.

We'll have much more to say about the electrical characteristics of transistors, but first, let's see how they're made.

HOW ARE SEMICONDUCTORS MANUFACTURED?

Semiconductor manufacture consists of three stages: material preparation, product assembly, and product testing.

Semiconductor manufacture can be broken into three basic steps. First, there is material *preparation,* which consists of producing pure monocrystalline (single-crystal) silicon (or germanium, or certain semiconducting compounds), properly doped as N-type and P-type in the desired regions. The second step is mechanical *assembly.* Tiny chips of appropriately doped crystal must have wire leads attached to them so they can be placed in the electrical circuitry of the system, and the device must be packaged to protect it from damage and contamination. And third, the product must be *tested* to assure its good performance.

Now, let's take a more detailed look at the manufacturing processes that turn out the great volume of semiconductor products required for today's electronic systems.

HOW IS THE MATERIAL PREPARED?

The first step of material preparation is to purify the raw semiconductor material.

Some semiconductor companies begin with trichlorosilane, a liquid chemical compound that contains a great deal of silicon which has been extracted from ordinary sand. Their first step is to purify this raw silicon, by chemical processing, into pure *poly*crystalline silicon metal ("polycrystalline" means "many different crystals," and even a single tiny piece of this material has many crystals, all jumbled up.) At TI in Dallas, for example, we have a chemical plant where we refine much of the world's supply of semiconductor grade silicon. Many semiconductor manufacturers buy silicon in this refined form, but it is still *poly*crystalline. So further processing is required.

After refining is complete, uniformly structured N- or P-type crystals are grown from the melted, purified silicon.

The next stage of material preparations is called "crystal growing," Fig. 7.4. The objective of this step is to produce *mono*crystalline silicon ("monocrystalline" means "one continuous orderly arrangement of atoms into a crystal). In crystal growing, a quantity of polycrystalline silicon is melted in a crucible. Then a small seed of monocrystalline silicon (about the size of a pencil eraser) is lowered into the crucible just far enough to touch the surface of the melted silicon. Since the seed is cooler, the molten silicon begins to crystallize on the seed, reproducing the crystalline structure of the seed. As it grows, it remains monocrystalline. The seed is slowly but continually lifted away from the melt, and more and more of the molten material accumulates to build up a large crystal. To keep growth uniform, the crystal and the pot are rotated in opposite directions as the crystal is pulled upward.

Figure 7-4. Crystal Growing

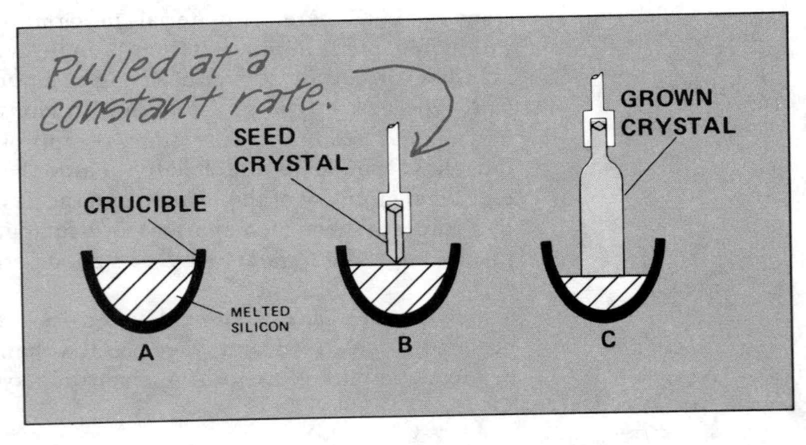

The finished product is a 2-6″ diameter cylinder, containing enough pure monocrystalline silicon to make millions of transistors or diodes when it is sawed up into tiny chips. In each of the chips, all of the atoms are lined up in exactly the right pattern, or crystal lattice, to produce a semiconductor.

PN junctions are created in one of three ways: by being grown, alloyed, or diffused.

Next, this pure silicon must be processed to have P-type and N-type material in just the right places. There are a number of ways to achieve this, but they fall into three categories: grown junction, alloyed junction, and diffused junction.

HOW DOES THE "GROWN JUNCTION" TECHNIQUE PRODUCE P-TYPE AND N-TYPE MATERIAL?

There are two ways of *growing* PN junctions: From molten silicon, or using vapor deposition (epitaxial). The molten method is not used much anymore, but it was the first production method used to make germanium transistors, and the first method for making silicon transistors. Picture the same set-up as in Fig. 7.4. Assume that the molten material in the crucible is already doped to produce N-type material. After some of the N-type crystal has been grown, a pellet of P-type impurity (about the size of a pin head) is dropped into the molten silicon. The pellet cancels out the effects of the N-type impurity, and makes the crystal begin to grow a region of P-type material. After a few seconds — and this is precisely timed — the P-region has grown about a half-thousandth of an inch thick, and a pellet of N-type materials is dropped in. This cancels out the P-type, and the crystal begins growing N-type again. The result is an NPN structure. It's much like a ham sandwich, with a thin layer of P-type material between two thick layers of N-type material. And like a ham sandwich being cut into tiny bridge sandwiches each with the same three layers, the grown junction crystal can be cut into many bars containing the same three layers.

The most common way to grow PN junctions is by "epitaxial growth" (the deposition of vapors onto slices of heated silicon).

Today, the much more popular method of growing a junction is called "epitaxial growth." (It means the crystal is usually grown onto a surface from a vapor.) In this technique, the first step is to grow a complete silicon crystal, doped N-type, for example. The bar is then sawed into wafer-thin discs, much as a loaf of bread is sliced. Each slice is ground flat and polished to mirror smoothness. Each of these N-type slices is placed on a heated platform in an enclosed chamber, as in Fig. 7.5. The slice is heated from below. Silicon

Figure 7-5.
Forming PN Junctions by Vapor Deposition

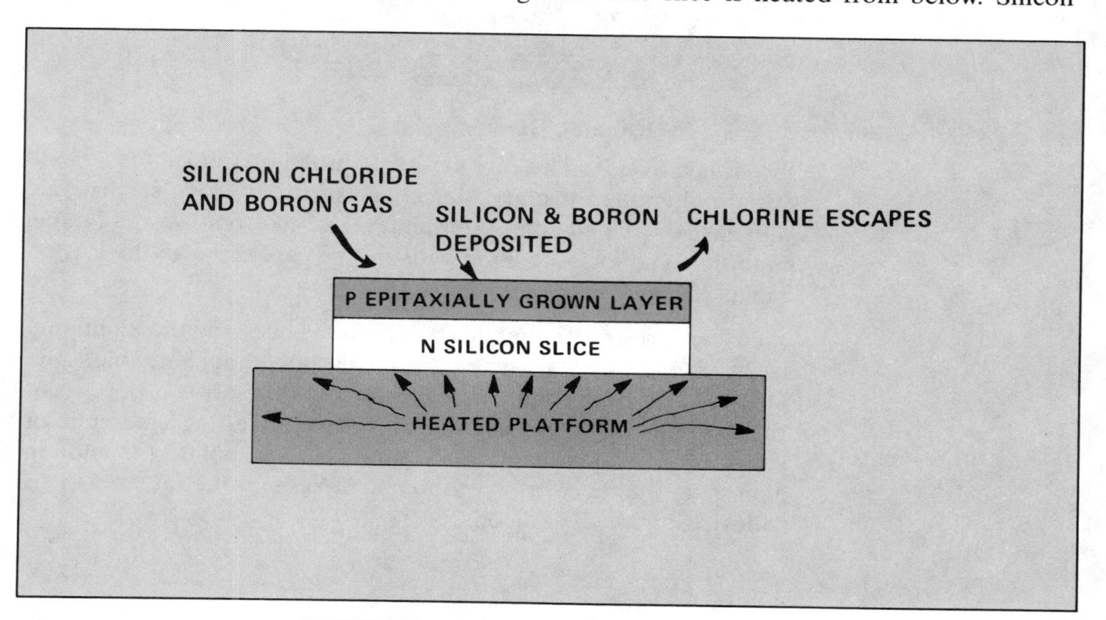

After the epitaxial layer is grown by vapor deposition to form a PN junction with the original slice, the remaining junction of the transistor is formed by diffusion.

chloride gas is pumped into the chamber. When the gas strikes the hot silicon slice, the gas decomposes and leaves monocrystalline silicon deposited on the slice. Since the gas in this case would carry a small amount of P-type impurity, such as boron, the epitaxially grown layer would be P-type. In this way, we have produced a PN junction. Most often, this is as far as epitaxial growth is carried, and the other N region is added by the diffusion technique, which we will study later.

HOW ARE JUNCTIONS PRODUCED BY THE ALLOY METHOD?

The alloy method of creating PN junctions has been used with silicon to make diodes and with germanium to make transistors. Bits of aluminum are melted into the surface of N-type semiconductor material.

The second basic method for producing semiconductor junctions is the alloy method. Although this method has been used to produce billions of *germanium* transistors and diodes, relatively few silicon devices are made this way, chiefly because of the problems that arise from the high melting point of silicon. Still, to avoid confusion, we will use an actual silicon alloy diode process for illustration.

In the alloy technique, the first step is to grow an N-type crystal. Then the crystal is cut into slices, and the slices are laid on a flat carrier, Fig. 7.6. A graphite disc of the same diameter as the

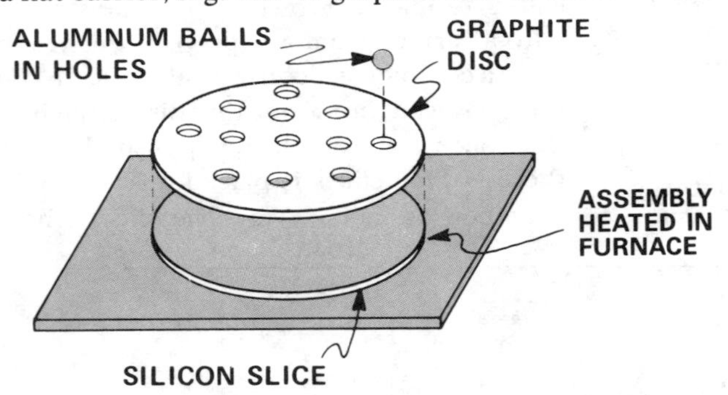

Figure 7-6. Forming PN Junctions with the Alloy Method

slice is laid over it. This disc has many little holes in it. A tiny ball of pure aluminum is placed into each of these holes, so that the balls rest on the silicon. Aluminum, just like boron, is a P-type impurity — so the aluminum balls in this case serve as the P-type dopant.

The carrier with the slice, the graphite disc, and the aluminum balls, is placed in a furnace which is hot enough to melt the aluminum. Some of the melted aluminum dissolves a portion of the silicon under it — just as a drop of water resting on a cube of sugar dissolves some of the sugar. At this point, the molten aluminum mixes with the silicon. We say the aluminum is "alloying" with the silicon.

The melted aluminum combines with the semiconductor material to create a P-type region next to the N-type silicon or germanium.

The assembly is then taken out of the furnace, and the aluminum hardens. This leaves pure aluminum on top, aluminum-silicon alloy in the middle (our P-region), and our original N-type silicon on the bottom.

Next, to avoid a short-circuit between the contact aluminum and the N-type silicon around the edge of the dot, the whole slice is dipped or washed in an acid which will dissolve silicon but not aluminum. This etches away the silicon surface except where it is protected by the aluminum ball. As shown in Fig. 7.7, this process

Figure 7-7. Cross-Section of Alloyed PN Junction

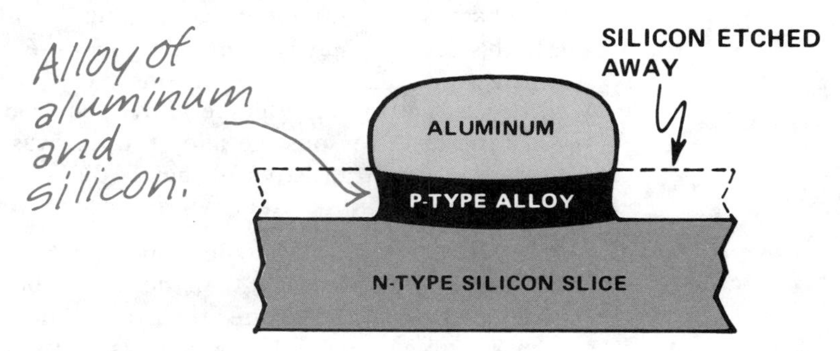

gives us a slice with many little P-type pedestals, each capped with aluminum.

The slice is then sawed to separate the little diode elements from each other. Each element is then mounted in a package to produce a finished device. Figure 7.8 shows this package, a glass

Figure 7-8. Silicon Alloy Diode

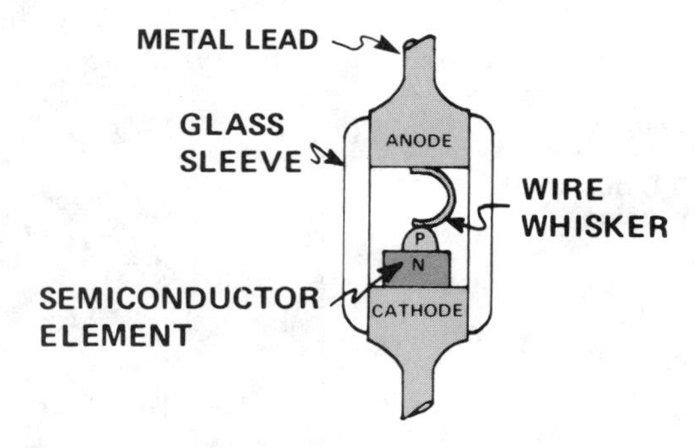

sleeve with cathode and anode leads stuck into the ends. The N side of the semiconductor element is attached to the cathode plug. Curving down from the anode lead is a tiny whisker of wire or metal ribbon that makes contact with the aluminum bump. This, then, is a silicon alloy diode.

It's easy enough to see how you could take the same slice before it is sawed apart, flop it over, and alloy dots on to the other side. This would produce PNP elements of the kind we use in transistors. Part of the fine art of making alloy transistors is to control the depth of the alloy from each side in such a way as to produce the very narrow base region we talked about. If the furnace temperature is a few degrees too low, or the time in the furnace a few seconds too short, we produce a base region so thick that it captures too many electrons. On the other hand, too high a temperature or too long a time, and the alloy "punches through" the base region altogether, contacting the alloy on the other side. In this case, we may have made a good fuse, but we haven't made a transistor. The serious problems involved in controlling factors like these at the high temperatures required to work with silicon, are the chief reasons the alloy process was never widely used to make silicon transistors — but it did work beautifully for germanium, and for certain types of silicon diodes.

In a transistor, incidentally, the ball we use to form the collector region is larger than the emitter ball, chiefly because free electrons from the emitter tend to radiate in all directions in the base — so a larger collector collects them more efficiently.

WHAT IS THE DIFFUSED JUNCTION PROCESS?

Nearly all integrated circuits are manufactured through the diffused junction process.

The diffused junction process is widely used to make transistors, and practically all integrated circuits are manufactured this way.

The diffused junction process also starts with a slice of monocrystalline silicon. In our example, Fig. 7.9, we use N-type

Figure 7-9. Forming PN Junctions by Diffusion—First Diffusion

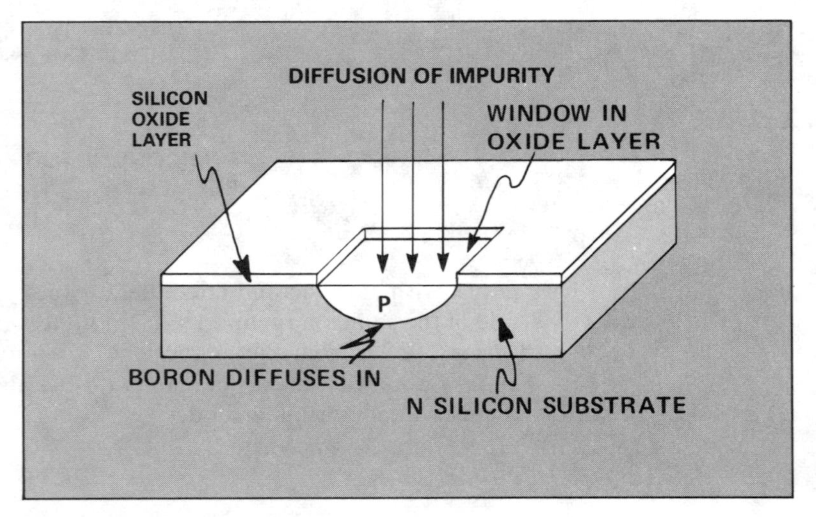

The diffusion process begins with a silicon "substrate" covered by a chemical-sensitive oxide layer and a light-sensitive photoresist layer.

silicon for the substrate ("substrate" means simply "underlayer"). The top of this substrate is made to oxidize, so that it is completely covered with a sealing layer of protective substance (silicon oxide). This oxide layer, in turn, is coated with a material called "photoresist." The photoresist is sensitive to light, something like a photographic film is. When a kind of master photographic negative called a "photomask" is placed over this photoresist, and light is shined on it, the light shows through selected areas of the mask, and causes chemical changes in those selected parts of the photoresist. Now, the slice can be washed and the selected parts of the photoresist will wash away, exposing the oxide layer in selected areas. And since the photoresist resists the attack of etchants, the slice can now be washed with etching chemicals to cut windows in the oxide layer, as shown in the figure.

Successive applications of light and chemicals create "windows" to the substrate, through which boron gas is diffused into the silicon to create P-type material. Then the substrate is covered once more, smaller windows are made over the previous ones, and the exposed silicon is treated with phosphorus to create N-type material.

Each slice has hundreds or thousands of such windows. If we are making typical small-signal transistors, each window is about twenty thousandths of an inch across. (If we are making integrated circuits, as we shall see in a later chapter, these windows may be even narrower than one half-thousandth of an inch across.)

The prepared slice, with the etched silicon oxide layer acting as a stencil, is then placed in a furnace and heated to a temperature near the melting point of silicon. Boron gas is pumped into the furnace, and it strikes the silicon where the windows have exposed it. The boron *diffuses* into the silicon crystal in the area of each window. This diffusion is a process of soaking in, just like water soaking into a sponge. The boron atoms take the places of a small proportion of the silicon atoms in the crystal structure; slowly, over several hours, they work their way down into the silicon crystal. Boron, of course, is a P-type dopant. So wherever it goes, the silicon becomes P-type. When the boron has diffused to the correct depth, making a layer of P-type silicon, the slice is removed from the furnace and the diffusion stops.

Now, we create a new layer of silicon oxide over the entire surface of the slice, windows and all. But again, we create new smaller windows in this layer. Each of *these* windows lies within the area of each old window. We put the slice back into the furnace, and this time expose it to gaseous *phosphorus*. The phosphorous atoms diffuse into the window areas and overpower the effects of the boron in a small region within the P-region,

Figure 7-10. Structure of a Transistor Formed by Successive Diffusions

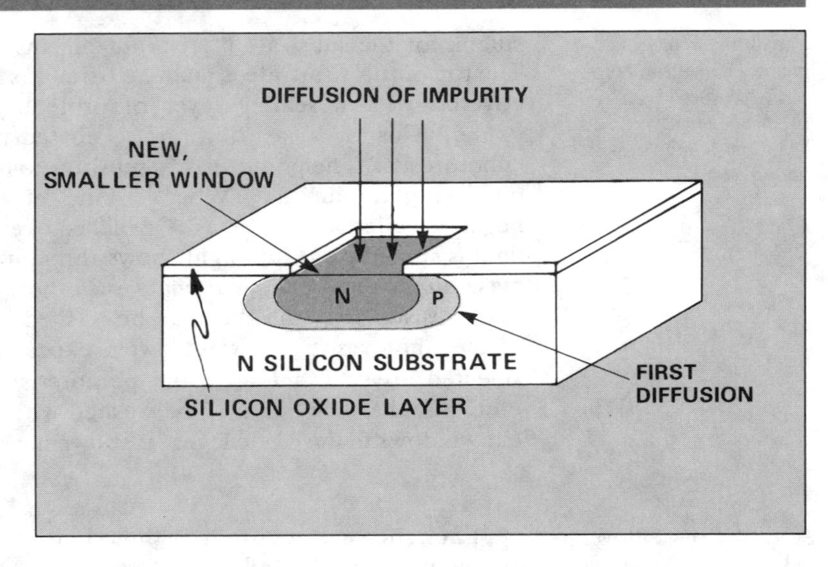

Cutting holes in the oxide to make contact with semiconductor material, coating the surface of each region with metal, and etching the metal to make metal pads makes it possible to attach wire leads to the device.

converting the silicon in this small area to N-type. Figure 7.10 shows this new N-region. It's easy to see the transistor structure now — a layer of P-type material between layers of N-type material.

In order to facilitate making contact with these layers by means of lead wires, we now plate the surface of each region with a thin layer of metal. This process is called "metallization." The process employs photomasking techniques of the sort we have just seen. As Fig. 7.11 shows, a thin layer of aluminum is plated over

Figure 7-11. Metallization and Contacts to Diffused Layers

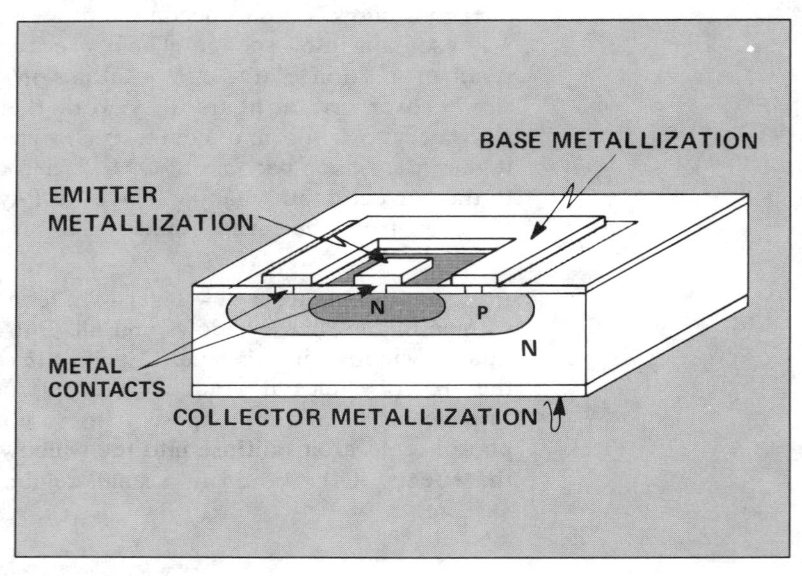

the emitter region and a similar layer is plated over the base region. The bottom of the slice is plated with a layer of gold. This provides not only a good electrical contact to the collector, but also a high-grade "solder" for attaching the chip intimately to the metal of the package. This makes for a strong physical bond, as well as a wide area through which heat can be dissipated to the package and to the outside world.

Large quantities of semiconductors can be economically produced with the diffusion process.

In actual manufacture, of course, several hundred of these transistor elements are formed on the same slice, which is then sawed apart after processing is complete. One of the chief advantages of the diffusion process is that it takes no more time or labor to process a whole slice full of transistors, than it would to process a single transistor.

The planar diffusion process produces semiconductors whose various regions and junctions lie in a single plane with the substrate surface. As a result, all operations that form devices are performed on the top of a slice.

WHAT ARE THE PLANAR AND MESA DIFFUSION PROCESSES?

The diffused transistor we have just described is an example of a *planar* diffused device. It is called "planar" because, as Fig. 7.12 suggests, all the regions (N, P, and N) and both of the junctions appear on the upper surface of the slice, in a *plane*.

Figure 7-12. Cross-Section of Transistors Formed Through Planar Diffusion

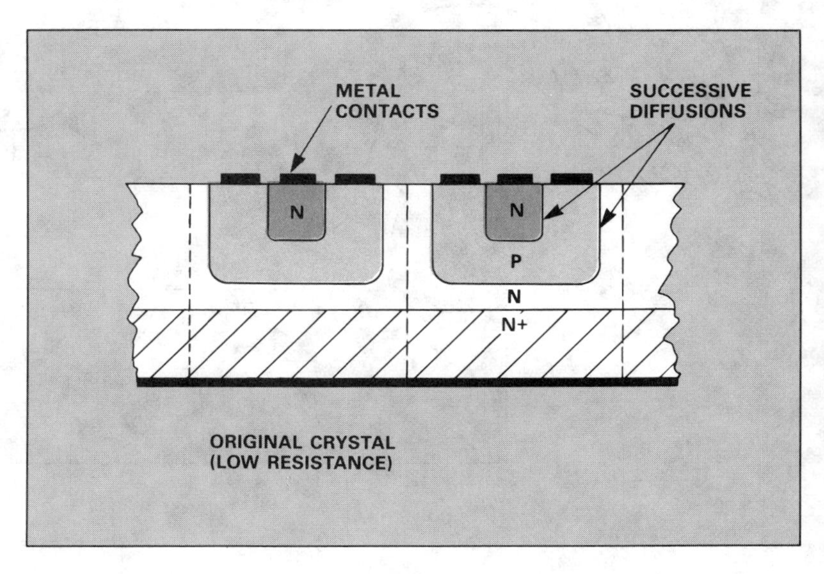

But there is another sub-category of the diffusion process called "mesa." We'll describe this shortly.

While we have Fig. 7.12 before us, however, we should mention a variation of the planar process, called "planar epitaxial." In this process, as shown in the figure, a lightly doped N-region can be grown epitaxially on top of a heavily doped substrate of N-type crystal (labeled N+). This process gives a lightly doped collector region which is quite desirable from the standpoint of high-voltage capability. It also produces a heavily doped low-resistance path for current to the collector contact.

In the mesa diffusion process, the P-type material is diffused over the entire substrate. Then the N-type material is diffused through windows. Finally, small valleys are etched between the devices to isolate the devices.

Now to the mesa process. Figure 7.13 shows how the process got its name — processing gives each element the shape of a tiny flat-topped mountain, or "mesa," as it's called in the Southwest. You recall that, in the planar process, the P-type base regions were diffused separately through windows in the silicon oxide; in the

Figure 7-13. Cross-Section of Transistors Formed Through Mesa Diffusion

Material is etched away to isolate devices.

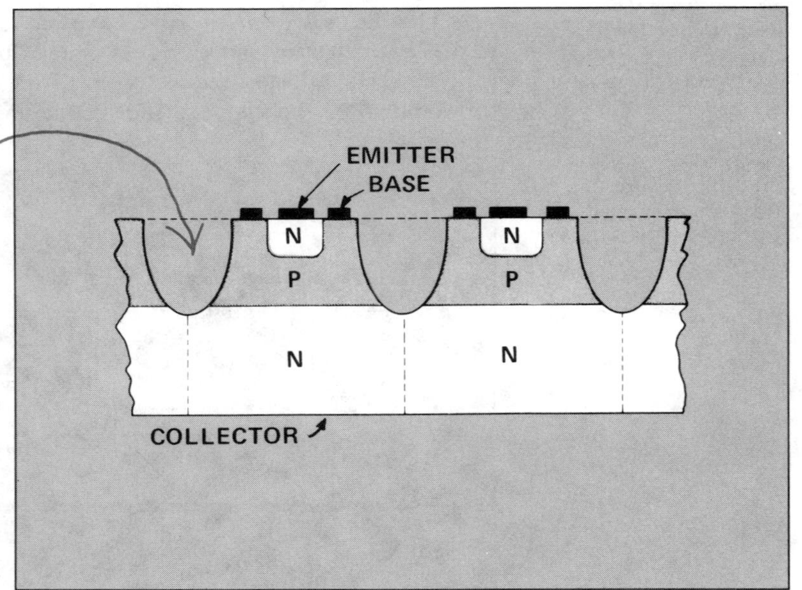

mesa process, on the other hand, a *single* P-region is diffused or deposited epitaxially over the *entire* surface of the slice. Then, N-type emitter regions are diffused into the slice individually, as in the planar process. Next, to prevent damage to the collector-base junctions when the chips are sawed apart, acid is used to etch away little valleys between devices. Then, the devices are sawed apart along the dotted lines extending down from the bottom of the valleys.

The mesa process produces fewer defects, but the planar process is cheaper.

One advantage of the mesa process over the planar is that it produces fewer defects in the slice. On the other hand, the planar process is inherently cheaper. We can't go into all the advantages and disadvantages of the various processes here. Suffice it to say that a semiconductor designer must make a great many complex trade-offs among all the factors of performance, reliability, and costs, when deciding the process steps he will use. In many cases, these decisions are so complex that designers must seek the aid of large computers to assist them in selecting the best possible combination of processes.

We have covered the three basic categories of material preparation methods: grown, alloy, and diffused. We have also discussed several important variations of these processes. We are primarily concerned with the planar diffused process for integrated circuit fabrication. The mesa diffused process is important in the production of high voltage transistors. The alloy process is no longer important in the manufacture of silicon devices.

Now, once the semiconductors have been created, they are still naked, defenseless, and without communication to the outside world. The next steps are the assembly processes that will make them into practical, usable devices. As an example, we'll take the assembly of diffused transistors.

HOW ARE DIFFUSED TRANSISTORS ASSEMBLED?

Completed semiconductor devices are packaged in metal cans or plastic blocks. Thin gold wires connect them to the outside leads.

Figure 7.14 shows how a diffused chip is assembled into a metal can. The chip is fastened to the gold-plated "header" platform by a heat process which alloys the chip to the gold plating. Typically, the platform itself serves as the collector lead and the platform is connected to one of the external leads. The emitter and base leads are insulated from the platform by glass, a portion of which is shown as the grey rings in the drawing.

Figure 7-14. Transistor Mounted in Metal Can

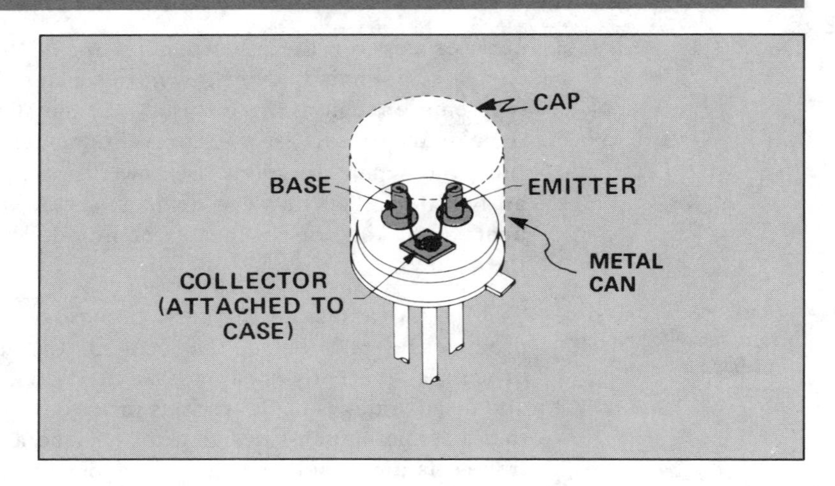

Thin gold wires — about the diameter of a human hair — are fused to the metallized areas that contact the emitter and base regions. The other ends of these wires are attached to the external leads, which go to the outside world. After these three connections are made, a protective cap is welded to the header. The transistor is now complete, and ready for testing.

An alternative way to package transistors is to encapsulate them in a solid block of plastic. Typically, three leads, flattened on the ends as shown in Fig. 7.15, are clamped in a temporary fixture.

The chip is alloyed to the center lead, and gold bond wires are attached to the other leads. This assembly is then placed in a mold, and plastic in a liquid state is injected into the mold. You can appreciate how rugged these transistors are, with every part embedded in a solid block of plastic.

Figure 7-15. Transistor and Lead Frame to be Sealed in Plastic

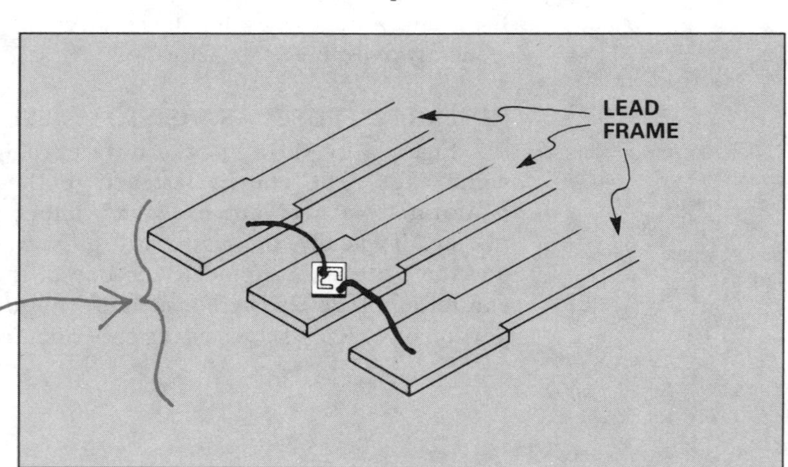

Contacts and chip are encased in plastic.

We have now covered the basic material preparation processes and the assembly processes that account for the manufacture of virtually all semiconductors. In bare outline form like this, the processes are probably much simpler than you imagined. But don't try to manufacture semiconductors at home in your own oven!

At this point, we have finished manufacturing the semiconductors, but they aren't ready for sale. Strange as it sounds, we still have to test each one to find out exactly what we've made!

WHY MUST EACH DEVICE BE TESTED FOR ELECTRICAL PERFORMANCE?

Since no two semiconductor devices are identical in all respects, each device must be tested after preparation and assembly.

In most modern industries, random *samples* of finished products are tested, and it's assumed that the *untested* products will function the same way. But in the semiconductor industry, *every* device must be tested or "characterized" before its type number can be determined. Despite some of the world's most sophisticated know-how, and experience accumulated over more than two decades, no semiconductor manufacturer can make a batch of devices that all turn out precisely alike in their electrical behavior.

The test results determine the category and type number assigned to each device.

So we must test each semiconductor by hooking its leads into test circuits. Depending on the results of these tests, devices from a single batch — subjected to the same processing throughout — may be divided into four or five different categories and sold under different type numbers. Of course, devices that do not meet specifications are rejected entirely.

Each device must meet certain different limits of measurement, as spelled out in its data sheet. Testing and characterization ensure that the device is within specified limits of forward voltage, reverse current, and many other characteristics that you are now familiar with, at least as far as diodes are concerned.

Although this testing *can* be carried out manually, using rather simple bench-top equipment, manufacturers use highly automated test systems like the TI-developed CAT machines (CAT stands for "computer automated testing"). These machines handle thousands of devices per hour, automatically testing and measuring many characteristics, analyzing the test results to determine the correct device type, and depositing the devices into their appropriate hoppers. Many devices are also subjected to operating-life tests, in which the transistors are actually operated day and night for weeks, often in high-temperature ovens, to ensure that they will perform reliably in severe environments.

We'll get into a few more manufacturing details in later chapters — in discussing integrated circuits, for example, we'll see how the planar process is employed to produce complicated complete circuits on a single chip or slice of silicon. In the next chapter, we begin to discuss transistor characteristics.

Quiz for Chapter 7

1. In a basic transistor circuit, the most significant factor which determines whether the transistor performs as a switch or an amplifier is:

- ☐ a. The type of transistor
- ☐ b. The control circuit which provides the input signal to the transistor
- ☐ c. The working circuit
- ☐ d. The transistor package
- ☐ e. None of the above

2. Switching and regulating action occurs in an NPN transistor:

- ☐ a. By withdrawing electrons from the transistor base, permitting a proportionally greater quantity of electrons to flow from the emitter to collector
- ☐ b. By increasing the number of electrons in the base region
- ☐ c. By replacing the buzzer in the circuit with a loudspeaker
- ☐ d. None of the above

3. Free electrons can flow in an NPN transistor from the emitter to collector with a good chance of evading capture by holes in the P-type base because:

- ☐ a. The P-type base is very narrow and lightly doped so that the holes are sparse and scattered
- ☐ b. Excess bound electrons that might repel the free electrons are being withdrawn from the base region by the control circuit
- ☐ c. Voltage pressure in the working circuit drives the free electrons across
- ☐ d. All of the above

4. The basic material in the manufacture of most of the semiconductor products presently being manufactured is:

- ☐ a. Monocrystalline silicon
- ☐ b. Grown-junction
- ☐ c. Gaseous boron
- ☐ d. N- and P-type aluminum hydroxide
- ☐ e. b and d above

5. Grown, alloyed, and diffused junctions are the three general techniques for:

- ☐ a. Mechanical assembly of semiconductor products
- ☐ b. Electrical testing of semiconductor products
- ☐ c. Producing P-type and N-type regions in the desired locations in the semiconductor element of any semiconductor device
- ☐ d. Application strictly to transistors
- ☐ e. a and b above

6. Epitaxial crystal growth (growing from a vapor rather than from a liquid) is mainly used to:

- ☐ a. Implement the alloy process
- ☐ b. Produce a lightly doped collector region on top of a more heavily doped original slice to improve the performance of diffused-junction devices
- ☐ c. Diffuse dopants into the slice through tiny windows in the oxide
- ☐ d. a and c above
- ☐ e. None of the above

7. A big advantage of the diffusion process is that:

- ☐ a. Individual hand-crafting of each separate semiconductor element is possible, thus achieving high quality
- ☐ b. It is a short, one-step process
- ☐ c. It takes no more labor or time to produce a whole slice full of hundreds of semiconductor elements than it does to produce only one transistor on a slice
- ☐ d. It doesn't require intricate, precise techniques
- ☐ e. b and d above

8. The very commonly used processes called "planar" and "mesa" are sub-categories of which method of producing P-type and N-type regions in semiconductor material?

- ☐ a. Grown junction
- ☐ b. Alloyed junction
- ☐ c. Diffused junction
- ☐ d. a and b above
- ☐ e. None of the above

9. The popularity of plastic-encapsulated semiconductor products is growing rapidly. From what was said about plastic transistors in this chapter, why should this be so?

- ☐ a. The plastic package contains a higher-quality chip
- ☐ b. The need for wire-bonding is eliminated
- ☐ c. The so-called plastic package contains a little metal package inside, and therefore provides the best possible seal against contamination
- ☐ d. A plastic device is much cheaper than a comparable metal-can device
- ☐ e. a and b above

10. After material preparation and mechanical assembly, the final electrical test step is necessary in semiconductor manufacture because:

- ☐ a. Unfortunately all the devices prepared and assembled the same way don't turn out exactly alike in their electrical behavior
- ☐ b. It goes without saying that it is desirable for all devices of a given type to be as uniform as possible in their electrical behavior
- ☐ c. Usually several different device types can be selected out of the semiconductors of one identical batch
- ☐ d. Devices that don't work at all must be thrown out entirely
- ☐ e. All of the above

THE PNP TRANSISTOR AND TRANSISTOR SPECIFICATIONS

Key Words

Breakdown Voltage
Conductivity
Gain
Leakage Current
Maximum Power Dissipation
Noise Figure
Operating Speed
PNP

Definitions are found in the glossary in the back of the book.

The PNP Transistor and Transistor Specifications

So far in our discussion of transistors, we have been talking about the NPN type exclusively. Now, we need to talk about the equally important PNP type, how they differ, and how the two types are often used to complement each other. Later in this chapter, we'll discuss specifications for both types.

HOW DOES A PNP TRANSISTOR DIFFER FROM AN NPN?

NPN and PNP transistors are exactly opposite in construction and operation—N-type is P-type and P-type is N-type.

As far as their *functions* are concerned, PNP and NPN transistors do precisely the same thing; they both switch and amplify electricity. They look alike and are manufactured by the same processes we have reviewed. But, much like British right-hand-steered cars and American left-hand-steered cars, they are exactly *opposite* each other in construction and in the way they operate.

Let's compare the two to see the ways in which they differ. First and most obvious, the roles of N-type and P-type material are *reversed*. P-type material is used as the base in the NPN transistor, whereas N-type material forms the base in the PNP type. By the same token, N-type material is used for emitter and collector regions of the NPN, but P-type is employed for emitter and collector in the PNP.

More significant, however, is the fact that the two types are precisely opposite in *operation*. Recall that, in the case of the NPN transistor, when we *withdraw* electrons from the P-type base, electrons can flow from emitter to collector and on through the working circuit. In contrast, with the PNP, if we put electrons *into* the base (the N-region), electrons will flow from *collector to emitter* and on through the circuit.

Current through a PNP transistor flows opposite the direction of current through an NPN transistor.

This difference in operation becomes even clearer when we compare the two schematic symbols, shown in Fig. 8.1. And here, it's best to think in terms of *conventional* current rather than electron current — in fact, throughout the rest of the book we will be using conventional current almost exclusively. Looking first at the NPN symbol, we see that it is controlled by a small current being pumped *into* the base, as *electrons* are *withdrawn*. The result is a large current flow *from* collector *to* emitter.

Figure 8-1. Current Flow Through NPN and PNP Transistors

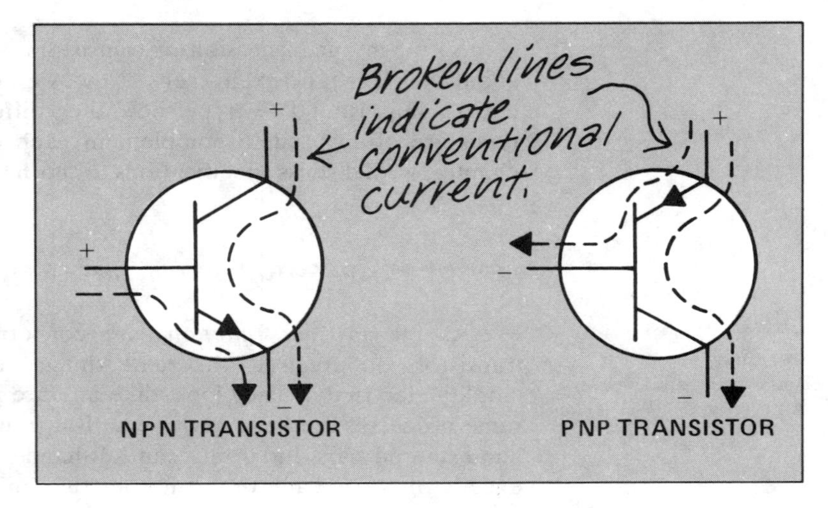

Notice now that the arrowhead is really a convenient feature of this symbol. It serves as a reminder of the direction of both the control current and the working current. The arrow points away from the base and away from the collector — signifying that conventional current must flow into the base and into the collector, but out of the emitter.

The arrowhead, which signifies a PN junction, points from the P region to the N region (the direction of conventional current).

When an arrowhead appears in *any* semiconductor symbol, it stands for a PN junction, and it points *from* P *to* N. Remember that the rule for PN junctions says that conventional current is *permitted* to pass from P to N, but is *blocked* when it attempts to pass in the reverse direction, from N to P. Therefore, current flows out the emitter of an NPN transistor.

Now consider the symbol for the PNP transistor. The arrowhead on the emitter symbol points *toward* the base instead of away from it. This is appropriate because the emitter is P-type and the base is N-type, and arrows always point from P to N. Here again, the arrow also tells us how current flows. In a PNP transistor, control current flows *from* emitter *to* base, in the direction of the arrow. Working current flows from emitter to collector.

You can see why we say the two transistor types are exactly equivalent but opposite in their operation. We need to qualify this a little, however. As a rule, NPN transistors can operate faster than PNP types — they can turn on and turn off quicker. Moreover, NPN transistors generally cost a little less to manufacture. As a result of these two factors, NPN types are much more widely used than PNP.

Although PNP transistors are somewhat slower and costlier than NPN transistors, they have important applications nonetheless.

In certain applications, each type has definite advantages over the other. So before we look at how PNP transistors work, let's find out why they're used. In spite of the fact that most PNP transistors are slightly slower and more expensive than NPN, they are useful and important products. Our example will show how the opposing characteristics of these two transistors can be put to good use by the equipment designer.

HOW ARE PNP TRANSISTORS USED?

PNP transistors are frequently used in conjunction with NPN transistors in amplifiers.

Figure 8.2 is the loudspeaker system we have used several times before. But now, we have improved it by using a PNP transistor.

In order to see how the addition of a PNP transistor improves the operation, we must review the operation of our original one-transistor system. Sound waves strike the microphone. In response, the microphone generates small surges or waves of current in the wire leading to the transistor base. The current in this wire is always flowing *into* the base, because we know that conventional current cannot flow out of the base of an NPN transistor — that would be opposite the direction of the emitter arrow.

Figure 8-2. Amplifier Circuit with NPN and PNP Transistors

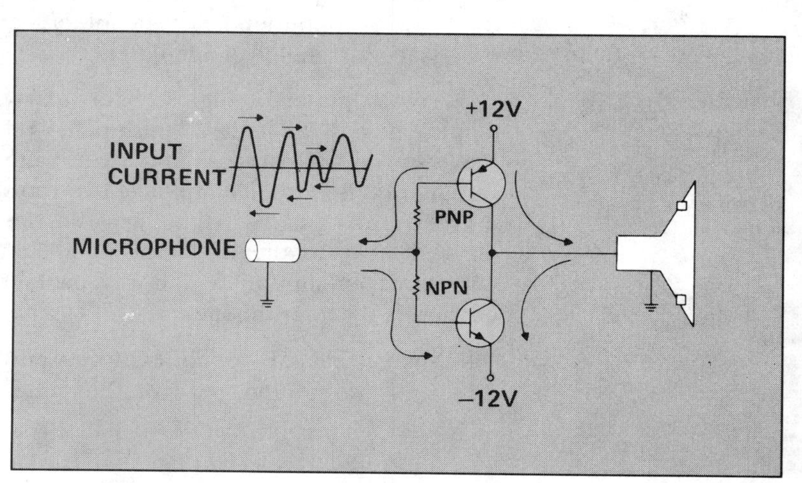

But remember that this direct current is not constant; it surges—first large, then small. The transistor, of course, amplifies these small surges by regulating the power-supply current to produce large surges flowing from collector to emitter. These large surges of current pass on to the loudspeaker. A surge of current makes the loudspeaker diaphragm jerk one way, creating a surge of air pressure. Then as the current decreases, the diaphragm *relaxes*, and springs back the other way. This pulls back on the air suddenly, generating a slight vacuum that follows the pressure wave. Thus, the loudspeaker creates an amplified version of the sound striking the microphone.

HOW WILL A PNP TRANSISTOR HELP?

Now, let's see what's wrong with this design, so we can understand how a PNP transistor will help. The trouble is that we used an extremely simple design for our original amplifier, for the sake of illustration. Making such a system work properly would require some very special capabilities in our microphone and loudspeaker. It would require some very *uncommon* devices that operate on surges of direct current — whereas standard microphones and loudspeakers require alternating current. Just as sound waves are alternating currents of air, so a typical sound-powered microphone responds by creating alternating electrical current, with current going first *into* the amplifier, and then *out* of it. Similarly, most standard loudspeakers require current going one way to push the diaphragm one way, and then the other way to *pull* the diaphragm the other way.

We *could* modify our original amplifier design by adding resistors and capacitors. This would make it a "class A" amplifier. But class A amplifiers waste a lot of power, making them impractical as high-power amplifiers.

Using NPN and PNP transistors in tandem creates a "push-pull" effect that simply and efficiently improves the amplifier's performance.

The amplifier of Fig. 8.2 is a "class B" amplifier. Class B amplifiers are also called "push-pull," referring to the fact that they use two transistors to provide an alternating output current. In our class B amplifier, we have two transistors. Now, a wave of current moving to the right turns on the NPN transistor, and a wave moving to the left turns on the PNP. Thus, the output current is an amplified and reversed copy of the alternating current in the microphone.

Such a class B amplifier, using both an NPN and a PNP transistor (rather than two of the same type) is called a *com-*

plementary class B amplifier, because the two transistors complement each other.

Two NPN transistors could be used instead, but the drive circuitry would be much more complicated and expensive. Using complementary transistors as we have, a simple pair of resistors will suffice for linking the transistors to the microphone. These resistors serve to bring the transistors to the threshold of conduction without turning on all the way.

There are many other such cases in which the availability of both NPN and PNP transistors enables designers to produce simpler, more economical designs that waste less power and perform better, than if only one kind of transistor were used.

HOW DOES A PNP TRANSISTOR WORK?

Most of the current carriers in a PNP transistor are positive holes. They perform the same function as the free electrons of an NPN transistor.

Figure 8.3 is a conceptual diagram of a PNP transistor. As in the NPN type, the base region of the PNP is extremely narrow — only a few ten-thousandths of an inch thick, and it is lightly doped. Since the base is N-type material, we will find a few free electrons in this region. In the drawing, the free electrons are represented by minus signs, and the positive holes in the P-regions are represented by circles. As we will see, the *holes* in the PNP perform the same function as free electrons in the NPN.

Just as we did with the NPN transistor, let's take that very brief moment in time when we start electricity flowing in the working circuit, without doing anything at all to the base region. What happens? For a split fraction of a second, the P-type emitter emits positive holes across the base. This occurs in just the same way that free *electrons* were emitted across the base in the NPN

Figure 8-3. PNP Transistor Cross-Section

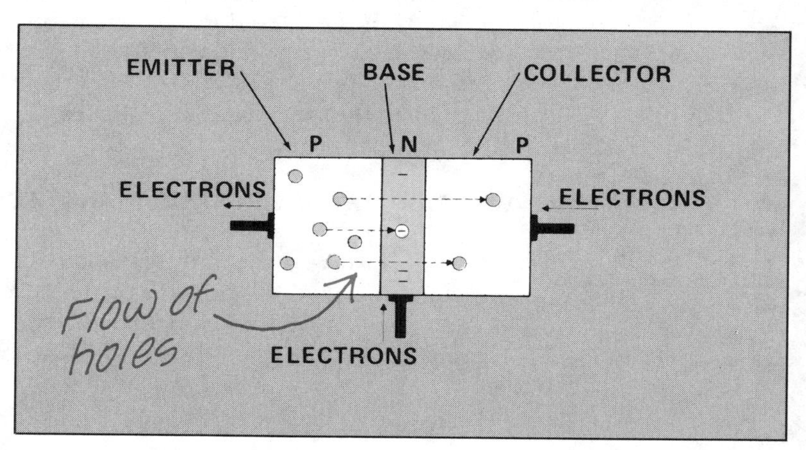

transistor. Since the base is narrow and lightly doped, a great many holes make it across the base — but just for an instant. Because during this time, some of the holes are getting stuck in the base region by meeting the free electrons. Every positive hole that gets caught in the base region by having an electron fall into it will tend to raise the positive charge in the base. Finally, the positive charge built up in this N-region exerts enough repelling force to stop further migration of holes. Once no more holes can enter the N-region, conduction is halted.

In order to get conduction going again, the control circuit must restore the depleted supply of negative free electrons, by pumping in some more through the base lead. When this happens, the presence of new free electrons counteracts the increased positive charge. This diminishes the repelling force, and positive holes can start moving again. That's about all there is to it — the operation is very much like that of the NPN transistor, except exactly opposite.

You may still have a bothersome question about the behavior of the PNP transistor. We've been talking about the movement of *positive holes* that cross junctions and flow through all three regions of the device. You may be asking yourself "do the holes continue on into the *wire*, providing some unusual kind of electricity?" That's a logical question, but holes can't do that; they move only through the *semiconductor* material. There's still just one kind of electricity — *electrons* in motion. If you are having a problem understanding the PNP, it may well be that you have heard so much about the positive holes that you've forgotten about electrons. "Holes moving" one way is simply a way of describing bound electrons moving the other way.

With this discussion of the PNP transistor, we've completed our study of the operation and applications of the most important devices in the semiconductor family, NPN and PNP transistors. Our next step is to analyze their data-sheet specifications, just as we studied the specifications of diodes.

Transistor performance can be described in terms of seven basic parameters: current gain, noise figure, maximum power dissipation, conductivity, leakage current, breakdown voltage, and operating speed.

WHAT ARE THE SEVEN BASIC TRANSISTOR SPECIFICATIONS?

We can take a more direct approach in studying transistor specifications than we did with diodes. In discussing diodes, we first had to consider diode behavior, by analyzing the graph of current versus voltage. But in the case of transistors, we have already covered the essentials of their behavior — we already know how the *control* current *switches and regulates* the *working* current.

A transistor data sheet contains a great deal of information. But as it turns out, the most important information can be spelled out in just seven major specifications. So we're going to go through the seven, one at a time:

Current gain, a measure of amplification, is the ratio of working current to control current.

Current gain (h_{FE} or h_{fe} or beta) is probably the best-known transistor specification. It tells us the number of times the transistor *multiplies* the control current to produce the working current. For example, if one milliamp of control current withdrawn from the base results in 100 milliamps of working current flowing out the collector, the current gain would be 100. Synonyms for current gain are "beta" and "common-emitter forward current transfer ratio." The symbol for current gain is written two different ways. When it's printed "h_{fe}," this is the "small-signal" current gain, and refers to the waves. It's possible for the *total* current to be amplified by a different amount than the *waves* in the current. That's why you may see both static and small-signal gain on the same data sheet.

Noise figure is a measure of the inherent distortion present in the transistor's output current.

Noise figure (NF) is the second basic transistor specification. Noise in an electrical circuit is any *unwanted* signal. This noise becomes audible in systems that produce sound — the random hissing, crackling sound you hear in your radio or phonograph is noise made audible. Every electrical device produces noise to some degree; it's caused by random, unwanted fluctuations in current and voltage that occur as a result of electrons flowing in a turbulent, uneven fashion. The output current of a transistor is not a perfectly amplified copy of the input current, but is slightly ragged. The noise figure of the transistor is an indication of *how* ragged and imperfect the working current is. Noise is expressed in units of relative *loudness* called "decibels."

The remaining five important transistor specifications are easy to understand because they are essentially the same as the five diode specifications we have already studied. In fact, these same five specifications appear, with only slight modifications, in the data sheet of almost every semiconductor device. They are truly *universal* semiconductor specifications. Moreover, they apply not just to semiconductors, but to all switching and regulating devices — including valves in water pipes.

Figure 8.4 lists the first two transistor specifications we have already covered, as well as the five universal specifications. As the

Figure 8-4. Basic Transistor Specifications

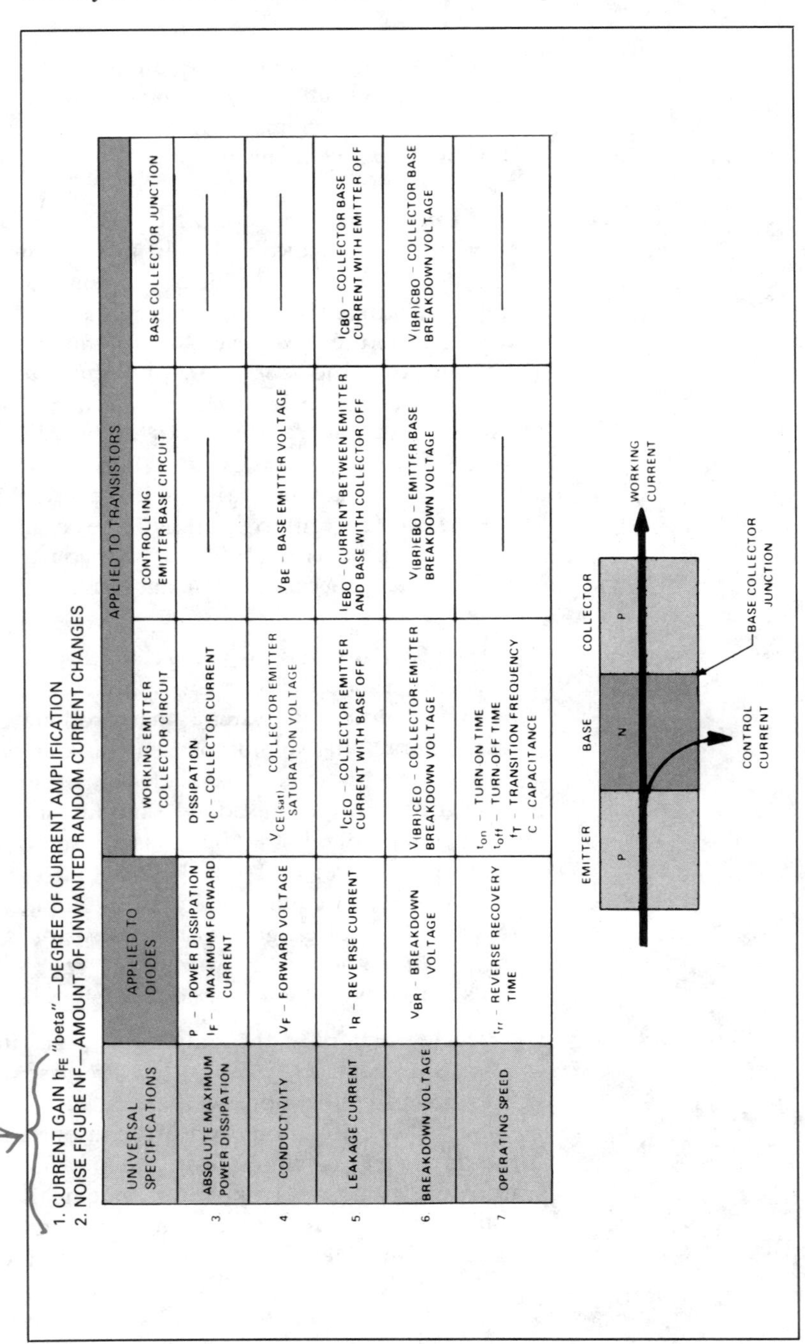

1. CURRENT GAIN h_{FE} "beta"—DEGREE OF CURRENT AMPLIFICATION
2. NOISE FIGURE NF—AMOUNT OF UNWANTED RANDOM CURRENT CHANGES

	UNIVERSAL SPECIFICATIONS	APPLIED TO DIODES	APPLIED TO TRANSISTORS		
			WORKING EMITTER COLLECTOR CIRCUIT	CONTROLLING EMITTER BASE CIRCUIT	BASE COLLECTOR JUNCTION
3	ABSOLUTE MAXIMUM POWER DISSIPATION	P – POWER DISSIPATION I_F – MAXIMUM FORWARD CURRENT	DISSIPATION I_C – COLLECTOR CURRENT	—	—
4	CONDUCTIVITY	V_F – FORWARD VOLTAGE	$V_{CE(sat)}$ COLLECTOR EMITTER SATURATION VOLTAGE	V_{BE} – BASE EMITTER VOLTAGE	—
5	LEAKAGE CURRENT	I_R – REVERSE CURRENT	I_{CEO} – COLLECTOR EMITTER CURRENT WITH BASE OFF	I_{EBO} – CURRENT BETWEEN EMITTER AND BASE WITH COLLECTOR OFF	I_{CBO} – COLLECTOR BASE CURRENT WITH EMITTER OFF
6	BREAKDOWN VOLTAGE	V_{BR} – BREAKDOWN VOLTAGE	$V_{(BR)CEO}$ – COLLECTOR EMITTER BREAKDOWN VOLTAGE	$V_{(BR)EBO}$ – EMITTFR BASE BREAKDOWN VOLTAGE	$V_{(BR)CBO}$ – COLLECTOR BASE BREAKDOWN VOLTAGE
7	OPERATING SPEED	t_{rr} – REVERSE RECOVERY TIME	t_{on} – TURN ON TIME t_{off} – TURN OFF TIME f_T – TRANSITION FREQUENCY C – CAPACITANCE	—	—

table shows, each of the five basic specifications appears on data sheets in more than one way; these variations are shown in the boxes. Although there is no need to memorize the meaning of all these symbols, you should learn to associate each specification with its significance in terms of one of the five universal specifications.

The polarity of current and voltage figures on data sheets is given in terms of conventional current: "positive in, negative out" and "positive higher, negative lower."

We will discuss each of the universal specifications in terms of a PNP transistor, rather than an NPN. Data sheets use *conventional* current — using a PNP, it is easy to remember that conventional current is emitted by the emitter, collected by the collector, and withdrawn from the base. Bear in mind, however, that the specifications of an NPN look just like those for a PNP, except that the direction of the currents and voltages is reversed. The direction is indicated by the *sign* of the value, a minus or plus. With current, the rule is "positive in, negative out." Collector current I_C, for example, is negative for a PNP because it is *leaving* the collector — but positive for an NPN because it is *entering* the collector. With voltage, the rule is "positive higher, negative lower," and it applies to the *first* of the two subscripts of the symbol. For example, with a PNP, the collector-emitter voltage difference V_{CE} is negative because the collector voltage (C is the first subscript) is lower than the emitter voltage. On the other hand, with an NPN, V_{CE} is positive, because the collector voltage is higher than the emitter voltage.

As Fig. 8.4 indicates, the universal specifications apply to each of the three different parts of the transistor. This is because there are three different paths that current takes through the device: The working current from emitter to collector, the control current from the emitter and out of the base, and the current that can flow — under certain conditions — from base to collector.

Maximum power dissipation is a measure of how much current-induced heat a transistor can tolerate without harm.

Maximum power dissipation (measured in watts) is power wasted by conversion to heat. The designer wants to know how much heat the transistor can stand before it malfunctions or burns up. Remember I_F in diodes? I_C is the corresponding specification for transistors. If the collector current exceeds this limit, destructively high temperatures will result. (Notice that the last two columns are left blank, because these values are negligible compared to the heat caused by the large working current, which is the collector current.)

The remaining four universal specifications are quite easy to remember because they can be illustrated by the functioning of a

Figure 8-5. 2N3250 Transistor Data Sheet

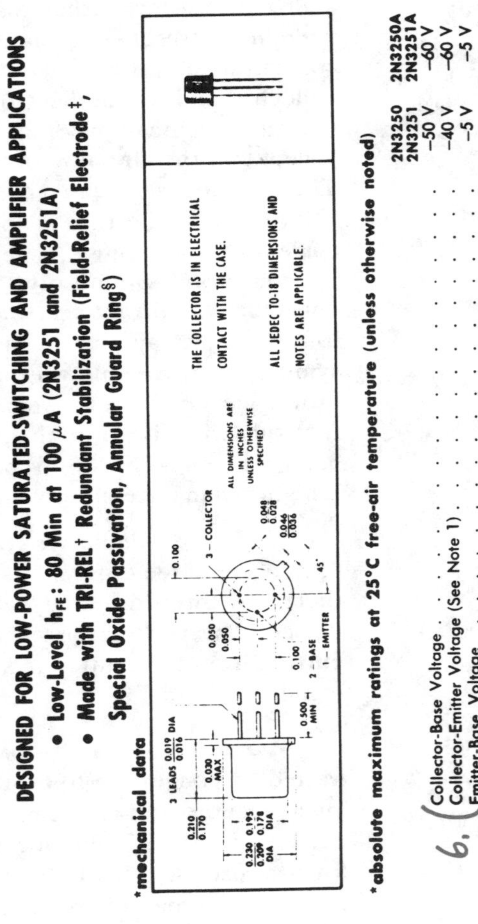

TYPES 2N3250, 2N3250A, 2N3251, 2N3251A
P-N-P EPITAXIAL PLANAR SILICON TRANSISTORS

TYPES 2N3250, 2N3250A, 2N3251, 2N3251A
BULLETIN NO. DL-S 679650, MARCH 1967
REPLACES BULLETIN NO. DL-S 657970, AUGUST 1965

DESIGNED FOR LOW-POWER SATURATED-SWITCHING AND AMPLIFIER APPLICATIONS

- **Low-Level h_{FE}: 80 Min at 100 μA (2N3251 and 2N3251A)**
- **Made with TRI-REL† Redundant Stabilization (Field-Relief Electrode‡, Special Oxide Passivation, Annular Guard Ring§)**

***mechanical data**

THE COLLECTOR IS IN ELECTRICAL CONTACT WITH THE CASE.

ALL JEDEC TO-18 DIMENSIONS AND NOTES ARE APPLICABLE.

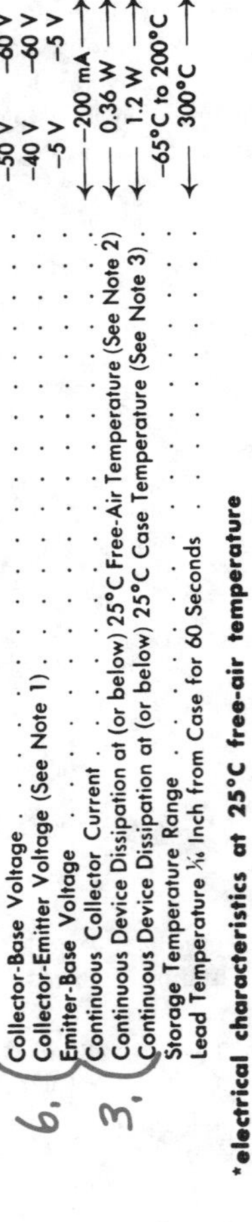

***absolute maximum ratings at 25°C free-air temperature (unless otherwise noted)**

	2N3250 2N3251	2N3250A 2N3251A
Collector-Base Voltage	−50 V	−60 V
Collector-Emitter Voltage (See Note 1)	−40 V	−60 V
Emitter-Base Voltage	−5 V	−5 V
Continuous Collector Current	←−200 mA→	
Continuous Device Dissipation at (or below) 25°C Free-Air Temperature (See Note 2)	← 0.36 W →	
Continuous Device Dissipation at (or below) 25°C Case Temperature (See Note 3)	← 1.2 W →	
Storage Temperature Range	−65°C to 200°C	
Lead Temperature 1/16 Inch from Case for 60 Seconds	← 300°C →	

***electrical characteristics at 25°C free-air temperature**

Figure 8-6.
2N3250 Data Sheet
(Continued)

PARAMETER		TEST CONDITIONS		2N3250		2N3250A		2N3251		2N3251A		UNIT
				MIN	MAX	MIN	MAX	MIN	MAX	MIN	MAX	
$V_{(BR)CBO}$	Collector-Base Breakdown Voltage	$I_C = -10\ \mu A$, $I_E = 0$		−50		−60		−50		−60		V
$V_{(BR)CEO}$	Collector-Emitter Breakdown Voltage	$I_C = -10$ mA, $I_B = 0$,	See Note 4	−40		−60		−40		−60		V
$V_{(BR)EBO}$	Emitter-Base Breakdown Voltage	$I_E = -10\ \mu A$, $I_C = 0$		−5		−5		−5		−5		V
I_{CEV}	Collector Cutoff Current	$V_{CE} = -40$ V, $V_{BE} = 3$ V			−20		−20		−20		−20	nA
I_{BEV}	Base Cutoff Current	$V_{CE} = -40$ V, $V_{BE} = 3$ V			50		50		50		50	nA
h_{FE}	Static Forward Current Transfer Ratio	$V_{CE} = -1$ V, $I_C = -0.1$ mA	See Note 4	40		40		80		80		
		$V_{CE} = -1$ V, $I_C = -1$ mA		45		45		90		90		
		$V_{CE} = -1$ V, $I_C = -10$ mA		50	150	50	150	100	300	100	300	
		$V_{CE} = -1$ V, $I_C = -50$ mA		15		15		30		30		
V_{BE}	Base-Emitter Voltage	$I_B = -1$ mA, $I_C = -10$ mA	See Note 4	−0.6	−0.9	−0.6	−0.9	−0.6	−0.9	−0.6	−0.9	V
		$I_B = -5$ mA, $I_C = -50$ mA			−1.2		−1.2		−1.2		−1.2	V
$V_{CE(sat)}$	Collector-Emitter Saturation Voltage	$I_B = -1$ mA, $I_C = -10$ mA			−0.25		−0.25		−0.25		−0.25	V
		$I_B = -5$ mA, $I_C = -50$ mA			−0.5		−0.5		−0.5		−0.5	V
h_{ie}	Small-Signal Common-Emitter Input Impedance	$V_{CE} = -10$ V, $I_C = -1$ mA, $f = 1$ kHz		1	6	1	6	2	12	2	12	kΩ
h_{fe}	Small-Signal Common-Emitter Forward Current Transfer Ratio			50	200	50	200	100	400	100	400	
h_{re}	Small-Signal Common-Emitter Reverse Voltage Transfer Ratio				10×10^{-4}		10×10^{-4}		20×10^{-4}		20×10^{-4}	
h_{oe}	Small-Signal Common-Emitter Output Admittance			4	40	4	40	10	60	10	60	µmho

NOTES:
1. These values apply between 0 and 200 mA collector current when the base-emitter diode is open-circuited.
2. Derate linearly to 200°C free-air temperature at the rate of 2.06 mW/deg.
3. Derate linearly to 200°C case temperature at the rate of 6.9 mW/deg.
4. These parameters must be measured using pulse techniques. $t_p = 300\ \mu s$, duty cycle $\leq 2\%$.

† Trademark of Texas Instruments
‡ Patent Pending
§ Patented by Texas Instruments
*Indicates JEDEC registered data

Figure 8-7. 2N3250 Data Sheet (Continued)

TYPES 2N3250, 2N3250A, 2N3251, 2N3251A
P-N-P EPITAXIAL PLANAR SILICON TRANSISTORS

***electrical characteristics at 25°C free-air temperature (continued)**

PARAMETER		TEST CONDITIONS	2N3250 2N3250A		2N3251 2N3251A		UNIT
			MIN	MAX	MIN	MAX	
$\lvert h_{fe} \rvert$	Small-Signal Common-Emitter Forward Current Transfer Ratio	$V_{CE} = -20$ V, $I_C = -10$ mA, $f = 100$ MHz	2.5		3		
f_T	Transition Frequency	$V_{CE} = -20$ V, $I_C = -10$ mA, See Note 5	250		300		MHz
C_{obo}	Common-Base Open-Circuit Output Capacitance	$V_{CB} = -10$ V, $I_E = 0$, $f = 100$ kHz		6		6	pF
C_{ibo}	Common-Base Open-Circuit Input Capacitance	$V_{EB} = -1$ V, $I_C = 0$, $f = 100$ kHz		8		8	pF
$r_b'C_c$	Collector-Base Time Constant	$V_{CE} = -20$ V, $I_C = -10$ mA, $f = 31.8$ MHz		250		250	ps

NOTE 5: To obtain f_T, the $\lvert h_{fe} \rvert$ response with frequency is extrapolated at the rate of −6 dB per octave from f = 100 MHz to the frequency at which $\lvert h_{fe} \rvert$ = 1.

***operating characteristics at 25°C free-air temperature**

PARAMETER		TEST CONDITIONS	2N3250 2N3250A	2N3251 2N3251A	UNIT
			MAX	MAX	
NF	Spot Noise Figure	$V_{CE} = -5$ V, $I_C = -100$ μA, $R_G = 1$ kΩ, $f = 100$ Hz	6	6	dB

***switching characteristics at 25°C free-air temperature**

PARAMETER		TEST CONDITIONS†	2N3250 2N3250A	2N3251 2N3251A	UNIT
			MAX	MAX	
t_d	Delay Time	$I_C = -10$ mA, $I_{B(1)} = -1$ mA, $V_{BE(off)} = 0.5$ V, $R_L = 275$ Ω, See Figure 1	35	35	ns
t_r	Rise Time		35	35	ns
t_s	Storage Time	$I_C = -10$ mA, $I_{B(1)} = -1$ mA, $I_{B(2)} = 1$ mA, $R_L = 275$ Ω, See Figure 2	175	200	ns
t_f	Fall Time		50	50	ns

**Figure 8-8.
2N3250 Data Sheet
(Continued)**

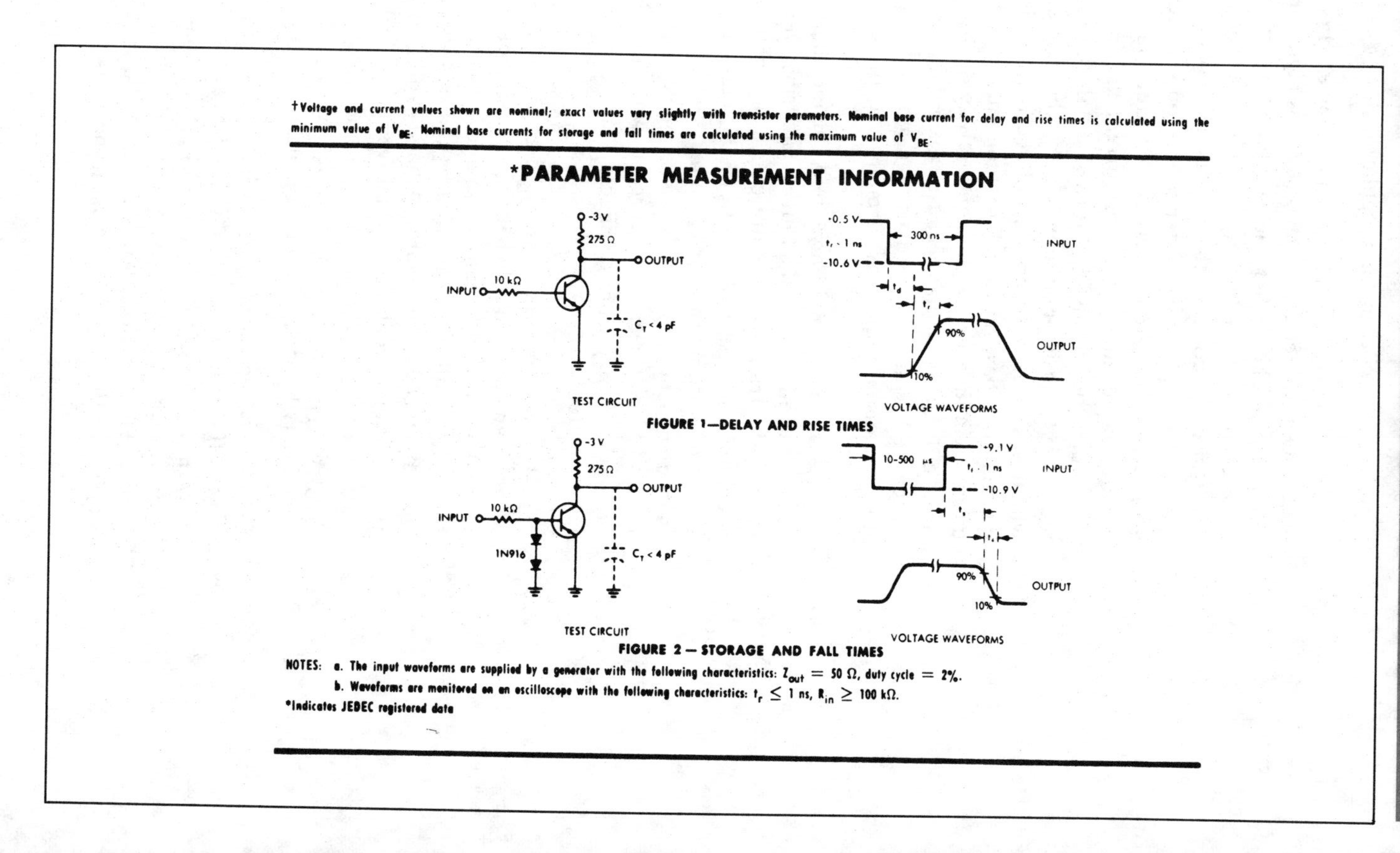

very familiar switching and regulating device — a water faucet.

Conductivity is a measure of how easily current flows through the working and control circuits.

Conductivity is an indication of how easily current flows when the device is turned on. A big faucet has high conductivity — water flows freely. The higher the conductivity, the lower the forward voltage drop required to move a curtain amount of current. In the first transistor column, the conductivity of the working circuit is specified as "collector-emitter saturation voltage" V_{CE} (sat), which is the voltage difference between emitter and collector required to maintain a certain amount of current moving in the *working* circuit, when the transistor is turned on all the way ("saturated"). In the second transistor column, we find the base-emitter voltage V_{BE} which is simply the forward voltage of the diode formed by the emitter and base required to maintain a certain current in the circuit — a measure of the conductivity of the *control* circuit. The third column is blank because, except in rare cases, no one is concerned with a forward voltage across the base-collector junction, because this diode is normally not conducting forward.

Leakage current is the small current that still flows across a PN junction when a transistor or diode has the junction reverse-biased.

Leakage current is unwanted current that flows when it ought to be blocked. Think of a faucet that drips even though it is turned off. In the transistor, we are concerned with leakage in all three specification areas. (The letter "O" stands for "open circuit." It simply means that the third terminal — the one that is not involved in the measurement — is disconnected. For example, I_{CEO} means collector-emitter current with the *base* unconnected. Similarly, I_{CBO} means collector-base current with the *emitter* unconnected.)

Breakdown voltage is the highest reverse voltage that a transistor or diode PN junction can withstand without damage.

Breakdown voltage is the highest voltage pressure that the device can hold back when it is turned off. If this limit is exceeded, the device breaks down and lets a large amount of current through. Think of a water faucet that is turned off, being broken open by excessive pressure. A handy thing to remember is that a transistor can break down wherever it can leak — in practice, this means through all three paths. In the third column, for example, $V_{(BR)CBO}$ means the breakdown voltage of the base-collector junction (when the emitter is open).

Operating speed is a measure of how quickly a transistor's output working current responds to changes in the input control current.

Operating speed refers to how quickly the transistor can turn on or off ("switching speed"), or change from one level of working current to another ("regulating" or "amplifying" speed). In water faucet terms, this is the time required to turn the handle back and forth. This specification is listed in the first column, because it applies to the speed with which the *working current* responds to changes in control current. In the case of *switching* transistors, we speak of the time to turn on the device, usually to 90 % of full voltage; or to turn it off, usually to 10 % of full voltage. Or we speak of "rise time" and "fall time." In any case, we're just talking about the operating speed of the device. On the other hand, *amplifying* transistors are given a frequency specification instead of switching times. The most common is the *transition frequency* f_T. In effect, this is the highest frequency at which the transistor can operate usefully as an amplifier. In certain cases, neither switching times nor frequencies are listed. In such cases, *other* specifications will show how well the device works at frequences that are spelled out in the test conditions.

If you're a conscientious reader, you're probably worried by now. You probably have a nagging feeling that we haven't told you all about how these specifications are measured — or you feel you don't understand what we have told you. It may make you feel more confident to know that there are more than *one hundred fifty* standard semiconductor specifications, and a lot of special ones — and very few practicing engineers know them all!

Take heart — if you have a good grasp of the five universal semiconductor specifications we have shown you, then you're well-equipped to puzzle out the meaning of almost any specification. To give you some practice, we have reproduced a complete 2N3250 Series data sheet. Even though this device may not be manufactured anymore, it serves to illustrate the type specifications included on data sheets. We have keyed the important specifications on the data sheet to Fig. 8.4 by the numbers 1 through 7 and the brackets to the left of the specifications.

In the next chapter, we'll explain two relatively simple but very interesting kinds of devices: thyristors, and optoelectronic semiconductors. If you have heard of field-effect transistors (FETs) and wonder why we haven't included them yet, it's because they differ in operation from the NPN and PNP transistors, and because their most important use is in integrated circuits. So we'll discuss them when we get into our discussion of integrated circuits.

Quiz for Chapter 8

1. The most important performance feature that distinguishes a PNP transistor from an NPN type is:

- ☐ a. The PNP is turned on by conventional current being withdrawn from the base, while the NPN is turned on by conventional current being forced into the base.
- ☐ b. Nothing. The two perform identically in every respect.
- ☐ c. The NPN switches electricity and the PNP regulates electricity
- ☐ d. The PNP works better than the NPN
- ☐ e. c and d above

2. Arrowheads in semiconductor symbols represent PN junctions and point from P to N in the direction that conventional current is permitted to pass. With this in mind, which symbol is the PNP transistor?

- ☐ a. 
- ☐ b. 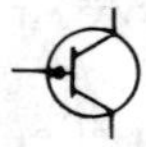
- ☐ c. 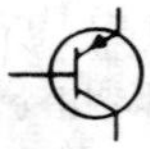
- ☐ d. 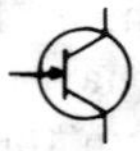
- ☐ e.

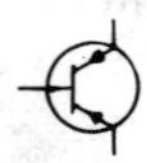

3. What is the advantage of using an NPN and a PNP transistor instead of two NPN transistors in a class B amplifier (that is, "push-pull"type — in which one transistor is on when the other is off)?

- ☐ a. The PNP is cheaper and works just as well.
- ☐ b. There is no particular advantage. An NPN can do the very same thing that the PNP does
- ☐ c. The PNP transistor can regulate alternating current passing through it
- ☐ d. The input can be connected fairly simply and directly to both transistor bases, because the PNP turns on when conventional current is withdrawn from the base while the NPN turns on when conventional current is pumped into the base
- ☐ e. None of the above

4. PNP transistors are less widely used than NPN types because:

- ☐ a. They were only recently invented, so designers are less familiar with them
- ☐ b. They perform an altogether different operation than NPN transistors — a function that happens to be considerably less useful
- ☐ c. They are generally slower in operating speed than NPN types
- ☐ d. They are generally more costly to manufacture than NPN types
- ☐ e. c and d above

5. In the PNP transistor, the working current consists of:

- ☐ a. Free electrons emitted by the collector and collected by the emitter
- ☐ b. Holes emitted by the emitter and collected by the collector. (Some holes get "trapped" in the base by combining with free electrons, and in order to maintain conduction, this positive charge must be neutralized by supplying electrons into the base lead)
- ☐ c. Holes pumped into the base terminal by the control circuit
- ☐ d. A combination of holes and free electrons traveling in the same direction
- ☐ e. a and c above

6. The h_{FE} or "beta" of a transistor is:

- ☐ a. The noise figure
- ☐ b. The current gain
- ☐ c. The ratio of output signal current to input signal current
- ☐ d. The degree of current amplification of the transistor
- ☐ e. All but a above

7. The significance of minus signs on certain specification values in data sheets (such as V_{CE} = -1 volt, and I_C = -10 milliamp) is:

- ☐ a. These specifications are in terms of electron current, which is negative
- ☐ b. All PNP specifications are negative, while all NPN specifications are positive
- ☐ c. If the specification is a current, such as I_C, a negative value means the current is flowing out of the particular terminal at which the measurement is made — in this case, the collector
- ☐ d. If the specification is a voltage, such as V_{CE}, the minus sign means the voltage pressure at the terminal indicated by the first subscript (here, the collector) is lower than the voltage pressure at the terminal indicated by the second subscript (here, the emitter)
- ☐ e. c and d above. (In other words, the sign indicates direction: Positive in, negative out — and positive higher, negative lower)

8. Some of the five universal specifications of a transistor are subdivided into separate specifications for:

- ☐ a. The control circuit (emitter base junction), the working circuit (emitter collector path,) and the base collector junction
- ☐ b. Switching and regulating
- ☐ c. Information and work
- ☐ d. Current and voltage

9. Which one of the five universal semiconductor specifications (listed below) describes how easily a device passes current (how much voltage difference is required to move a certain amount of current)?

- ☐ a. Absolute maximum power dissipation
- ☐ b. Conductivity
- ☐ c. Leakage current
- ☐ d. Breakdown voltage
- ☐ e. Operating speed

10. If neither switching times nor maximum frequencies are listed on a data sheet, what is another indication of operating speed that will usually be specified?

- ☐ a. Noise figure
- ☐ b. Conductivity
- ☐ c. Test-condition frequencies for other parameters
- ☐ d. Collector current
- ☐ e. None of the above

9 THYRISTORS AND OPTOELECTRONICS

Key Words

Light-Emitting Diode (LED)
Phase Control
Photodiode
Phototransistor
SCR
Thyristor
Triac

Definitions are found in the glossary in the back of the book.

Thyristors and Optoelectronics

This chapter completes our study of the important discrete semiconductor devices, by introducing thyristors and optoelectronic semiconductors. These devices are outgrowths of the basic diode and transistor technology we have already covered. So as we describe these devices, you'll find most of the concepts already familiar. You'll be putting to use almost everything you've already learned about diodes and transistors. This chapter serves as a graphic demonstration that, once you've mastered the fundamentals of semiconductor theory, variations and applications of it are relatively easy to understand.

WHAT ARE THYRISTORS?

Thyristors are high-power semiconductor switches that need no control current after they are turned on. The two most important types of thyristors are silicon controlled rectifiers (SCRs) and triacs.

Thyristors are switching devices that don't require any control current once they are turned on. All they require to snap them on is a quick pulse of control current. When the pulse current stops, thyristors keep going as though nothing happened. As you might imagine, this is a very useful kind of *switch*.

There are several kinds of thyristors, but we will discuss only the two most important types — silicon controlled rectifiers (SCRs), and triacs. Like transistors, these devices have two terminals for working current and one terminal for control current. But unlike transistors, thyristors don't require any further control current once they are turned on. Consequently, the control circuitry is typically quite simple and consumes little power. This is the chief advantage of thyristors over transistors.

Thyristors are mostly used in the "act" stage of systems, to control power going to a working device such as a motor. As you would suspect, they are generally high-power devices — they can handle a great deal of power without reaching harmful temperatures. Although they are switches, they're rarely used for processing information in the "decide" stage of systems, because their operating speed is typically much slower than that of transistors. Moreover, they typically can't be turned off by means of control current, so they can't be controlled continually as a transistor can. But what they do, they do so well that there's no substitute for them.

WHAT IS AN SCR?

An SCR normally blocks current in both directions. It admits forward current if it receives a gate pulse while forward voltage is being applied.

An SCR, as shown in Fig. 9. 1, is a semiconductor device that normally blocks conventional current attempting to pass *either* way between the anode and cathode. But when current is attempting to flow from anode to cathode, a quick pulse of current into the gate will turn the SCR on. And most important, the SCR *stays on,* even after the control gate current has stopped, as long as working current is being supplied. If the working current is turned off at some other point in the working circuit, then another gate pulse is required to restore conduction.

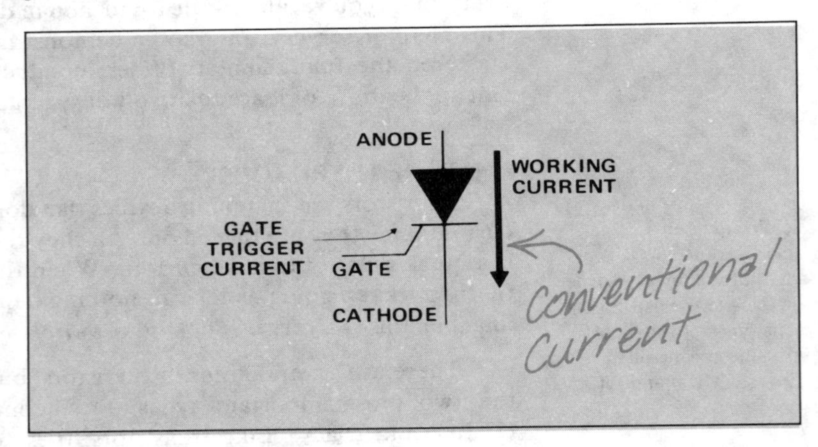

Figure 9-1. Silicon Controlled Rectifier (SCR)

Since the SCR normally does not permit any appreciable working current to pass in reverse, from cathode to anode, it functions like any other diode or rectifier. So it's a *rectifier* that can be *controlled.* Hence the name, "silicon controlled rectifier" or SCR. The formal name is "silicon reverse-blocking triode thyristor," but after many tongues got badly twisted, "SCR" became the popular name.

(This is a good place to point out that "diode," as we have seen, means *two* terminals. "Triode" means three. "Tetrode" means four. "Pentode" means five terminals. These terms mean the same thing as applied to vacuum tubes, and the functions of the tubes in many cases are similar to the functions of the semiconductors. Almost all transistors are triodes, of course, but a few transistor tetrodes are available.)

SCRs are commonly used in electronic automobile ignition systems. A small current from the breaker points provides the gate pulse.

Figure 9.2 shows the concept of a typical SCR application in a solid-state automobile ignition system. The power-supply circuit stores up electrical charge in the capacitor while the SCR is turned off. When the time comes for the spark plug to fire, the breaker points close for a moment, admitting a small control current to the gate. This triggers the SCR to the "on" state, discharging the large quantity of stored electricity from the capacitor through the few *primary turns* at the lower end of the spark coil diagram. This sudden surge of very high current generates a very high voltage in the secondary turns of the spark coil, and this voltage produces the spark. Quickly, when the charge on the capacitor has been depleted, the working current through the SCR stops. When the working current stops, the SCR turns off, permitting the capacitor to charge up again.

Figure 9-2.
SCR Used in Solid-State Automobile Ignition System

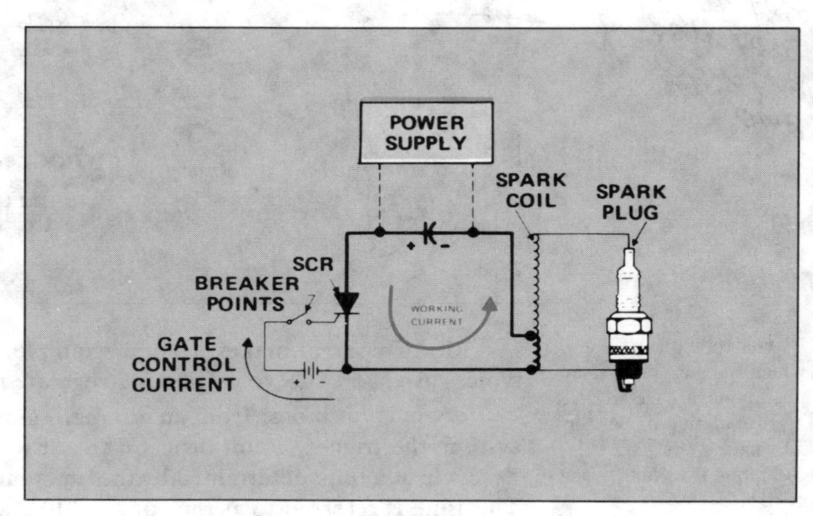

SCRs greatly reduce the wear on breaker points, while simultaneously permitting higher voltages and currents in the working circuit.

In ordinary ignition systems, the breaker points have to switch the full working current directly. As a result, breaker points wear out rapidly because of the high voltages and current they must switch — and this causes poor engine performance. The great advantage of the SCR in this application is that the breaker points have to handle only very low voltage and very little current. As a consequence, they last almost forever in good condition, and the working circuit can use higher voltages and current to give a hotter spark for better ignition.

Let's look at another typical application of the SCR. We have defined the SCR as a switching device. But strangely enough, it can also be used as a regulating device for alternating current. This is the basis for their wide-scale application in lamp dimmers, and as speed controls in food mixers, power tools, industrial motors, and many other regulating functions.

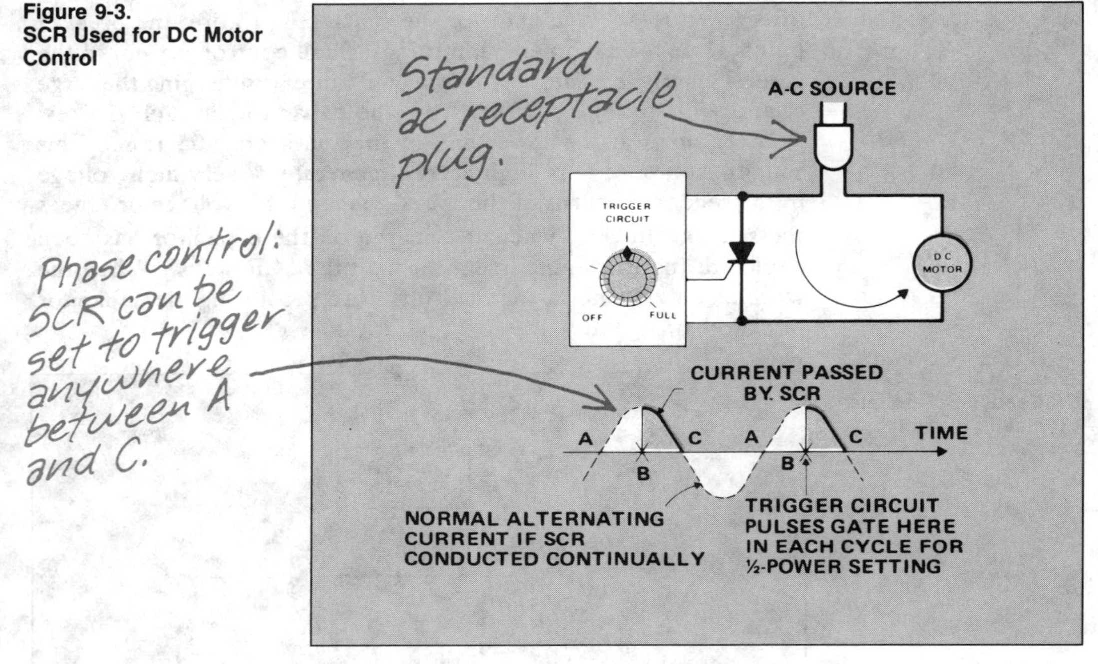

Figure 9-3. SCR Used for DC Motor Control

By controlling the time at which the gate pulse is applied, you can use an SCR as a regulating device for alternating current. This technique is called "phase control."

Motor control makes a good example. Figure 9.3 shows a typical use of an SCR to rectify and regulate the power supplied to a *direct*-current motor from an *alternating*-current source. In this system, the trigger circuit turns on the SCR once during each full ac cycle at a time determined by the manual control-knob setting. This time is referred to as the "phase," like a phase of the moon's cycle. The figure shows the waveform of working current as the ac source is attempting to pump it (the dashed line) and as the SCR admits it (solid line) when the control is set at *half*-power. At point A in each cycle, current tries to pass forward (from anode to cathode), but is blocked because the SCR is not yet turned on. At the phase of the cycle indicated by point B, the trigger circuit turns on the SCR, and current flows. At point C, the forward current has decreased to zero, so the SCR turns off again. The trigger-pulse timing can be varied to occur at any phase in the cycle between points A and C, thereby varying the average current supplied to the motor, across the entire range from zero to complete forward half-waves.

This method of power regulation is called "phase control." Its advantage is that virtually no power is wasted (dissipated) by being converted to heat. Heat is generated when we *regulate* current, not when we switch it. When we throttle current through a resistance, or some sort of regulating control element like a transistor or a rheostat, we make it fall through a voltage drop — and this always generates heat. But the SCR is always either fully on or off; it never throttles current part way.

Now that we know what the SCR does and where it's used, let's see how it works.

HOW DOES AN SCR WORK?

The semiconductor element of an SCR is a four-layer PNPN structure, as shown in Fig. 9.4a. Here's the key to how it works:

Figure 9-4. Structure and Operation of an SCR

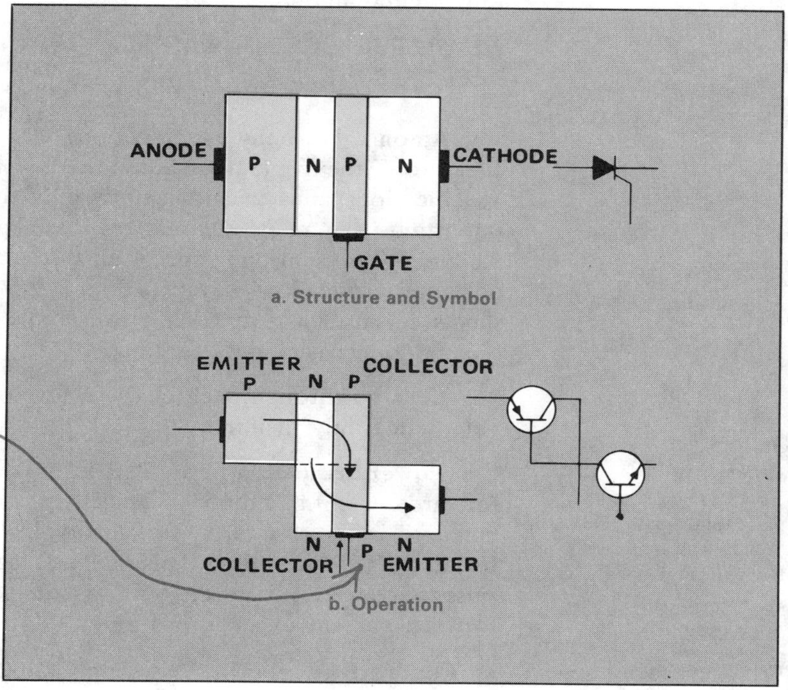

a. Structure and Symbol

b. Operation

Gate pulse triggers NPN side, which then turns PNP side on.

In its construction, an SCR resembles two transistors merged together. Applying a gate pulse turns one "transistor" on, which in turn causes the other to conduct.

Notice that if we take just the three regions counting from the left of the device, we have the equivalent of a PNP transistor. And if we take the three regions counting from the right, we have an NPN transistor. Figure 9.4b shows this "Siamese twin" arrangement. SCR's aren't made this way, but it's convenient to think of them as being divided into two transistor elements. The anode is connected to the P end of the stack, the cathode to the N side of the stack, and the gate to the other P-region.

Suppose that current is trying to flow from anode to cathode. It's blocked by both transistors. But if current is pumped into the NPN base by way of the gate terminal, the NPN transistor turns on — and *its working current* withdraws *control current* from the PNP base, turning *it* on. Even after the original gate current is stopped, therefore, the PNP working current continues flowing right into the base of the NPN, keeping it turned on. In this way, current continues to flow between anode and cathode until it is interrupted somewhere else in the circuit.

As you might imagine, it's possible to turn off the SCR by withdrawing current from the gate—but in most SCRs, this is not a very efficient operation. A few thyristors are designed for efficient gate turn-off — but these "gate-controlled switches" are rather uncommon.

WHAT SPECIFICATIONS ARE IMPORTANT ON AN SCR DATA SHEET?

Among the many specifications that appear on SCR data sheets, only the five that are based on the *universal* semiconductor specifications, plus two that apply only to thyristors, are really important. Figure 9.5 summarizes these specifications, and Figs. 9.6 and 9.7 present one page of an SCR data sheet, with the important specifications keyed to Fig. 9.5. As with the other data sheets, it is used as a representative sample even though the device type no longer is manufactured.

Let's go through each of these seven SCR specifications, to get them firmly in mind:

Power dissipation in an SCR will be exceeded if more than a specified "anode forward current" flows through the SCR.

Power dissipation in an SCR is the equivalent of maximum forward current in a diode. Typically, it is called "anode forward current," in the case of the SCR. A typical forward current rating for an SCR is two amps. This implies that power dissipation will raise the temperature of the SCR to a dangerously high level if the forward current exceeds two amps.

Conductivity is expressed by a "forward voltage" specification at a given current.

Conductivity of an SCR is simply an expression of how easily the device conducts electricity when it is turned on all the way, as was the case with transistors. You probably won't find the word "conductivity" on any data sheet, because it's meant to describe the concept only. The conductivity of an SCR is expressed as *forward voltage* V_F at a given current. A typical data sheet would list this as "static forward voltage" ("static" meaning with continuous direct current). This is the forward voltage pressure required to achieve a given level of forward current when the device is turned on. For an SCR, this maximum specification will always be somewhere in the neighborhood of 1.2 volts of forward voltage required. Why 1.2 volts? Look again at Fig. 9.4, the bar diagram of the SCR. Remember that we have two forward PN junctions. And remember that the forward voltage drop for a silicon PN junction is about 0.6 volt. Two times 0.6 volt is 1.2 volts, the typical forward voltage of an SCR.

SCRs block current in both directions. Leakage current, therefore, is expressed in two ways: as "anode reverse blocking current" and as "anode forward blocking current."

Leakage current, you'll recall, is the trickle of current that gets through when the device is supposed to be blocking. For a diode, we had only *reverse* leakage I_R. But since the SCR is a control device, it will block in *both* directions when not turned on. So in the SCR, we have *two* leakage specifications: One is "anode *reverse* blocking current." "Anode" just indicates where the current is measured. Reverse current, of course, is supposed to be

Figure 9-5. Basic SCR Specifications

If this current is exceeded, too much power is dissipated.

	"UNIVERSAL" SPECIFICATIONS	SCR SPECIFICATIONS	EXAMPLE VALUES
1.	POWER DISSIPATION	ANODE FORWARD CURRENT	ABSOLUTE MAXIMUM 2 A
2.	CONDUCTIVITY	V_F FORWARD VOLTAGE	MAXIMUM 1.2 V
3.	LEAKAGE CURRENT	I_R ANODE REVERSE BLOCKING CURRENT	MAXIMUM 1 μA
		I_F ANODE FORWARD BLOCKING CURRENT	
4.	BREAKDOWN VOLTAGE	REVERSE BLOCKING VOLTAGE	ABSOLUTE MAXIMUM 100 V
		FORWARD BLOCKING VOLTAGE	
5.	OPERATING SPEED	t_{on} TURN ON TIME	TYPICAL 1 μs
		t_{off} COMMUTATING TURN OFF TIME	TYPICAL 2 μs
6.		I_{GT} GATE TRIGGER CURRENT	MAXIMUM 100 μA
7.		V_{GT} GATE TRIGGER VOLTAGE	MAXIMUM 0.7 V

Figure 9-6. TIC106 Silicon Controlled Rectifier Data Sheet

SERIES TIC106

P-N-P-N SILICON REVERSE-BLOCKING TRIODE THYRISTORS

absolute maximum ratings over operating case temperature range (unless otherwise noted)

	TIC106Y	TIC106F	TIC106A	TIC106B	TIC106C	TIC106D	UNIT
Repetitive Peak Off-State Voltage, V_{DRM} (See Note 1)	30	50	100	200	300	400	V
Repetitive Peak Reverse Voltage, V_{RRM}	30	50	100	200	300	400	V
Continuous On-State Current at (or below) 80°C Case Temperature (See Note 2)	5						A
Average On-State Current (180° Conduction Angle) at (or below) 80°C Case Temperature (See Note 3)	3.2						A
Surge On-State Current (See Note 4)	30						A
Peak Positive Gate Current (Pulse Width ≤ 300 μs)	0.2						A
Peak Gate Power Dissipation (Pulse Width ≤ 300 μs)	1.3						W
Average Gate Power Dissipation (See Note 5)	0.3						W
Operating Case Temperature Range	–40 to 110						°C
Storage Temperature Range	–40 to 125						°C
Lead Temperature 1/16 Inch from Case for 10 Seconds	230						°C

NOTES:
1. These values apply when the gate-cathode resistance $R_{GK} = 1\ k\Omega$.
2. These values apply for continuous d-c operation with resistive load. Above 80°C derate according to Figure 3.
3. This value may be applied continuously under single-phase 60-Hz half-sine-wave operation with resistive load. Above 80°C derate according to Figure 3.
4. This value applies for one 60-Hz half-sine-wave when the device is operating at (or below) rated values of peak reverse voltage and on-state current. Surge may be repeated after the device has returned to original thermal equilibrium.
5. This value applies for a maximum averaging time of 16.6 ms.

Figure 9-7. TIC106 Data Sheet (Continued)

electrical characteristics at 25°C case temperature (unless otherwise noted)

PARAMETER		TEST CONDITIONS		MIN	TYP	MAX	UNIT
I_{DRM}	Repetitive Peak Off-State Current	V_D = Rated V_{DRM},	$R_{GK} = 1\ k\Omega$, $T_C = 110°C$			400	μA
I_{RRM}	Repetitive Peak Reverse Current	V_R = Rated V_{RRM},	$I_G = 0$, $T_C = 110°C$			1	mA
I_{GT}	Gate Trigger Current	$V_{AA} = 6\ V$,	$R_L = 100\ \Omega$, $t_{p(g)} \geq 20\ \mu s$		60	200	μA
V_{GT}	Gate Trigger Voltage	$V_{AA} = 6\ V$, $t_{p(g)} \geq 20\ \mu s$,	$R_L = 100\ \Omega$, $R_{GK} = 1\ k\Omega$, $T_C = -40°C$			1.2	V
		$V_{AA} = 6\ V$, $t_{p(g)} \geq 20\ \mu s$	$R_L = 100\ \Omega$, $R_{GK} = 1\ k\Omega$,	0.4	0.6	1	
		$V_{AA} = 6\ V$, $t_{p(g)} \geq 20\ \mu s$,	$R_L = 100\ \Omega$, $R_{GK} = 1\ k\Omega$, $T_C = 110°C$	0.2			
I_H	Holding Current	$V_{AA} = 6\ V$, $T_C = -40°C$	$R_{GK} = 1\ k\Omega$, Initiating $I_T = 10\ mA$,			8	mA
		$V_{AA} = 6\ V$,	$R_{GK} = 1\ k\Omega$, Initiating $I_T = 10\ mA$			5	
V_{TM}	Peak On-State Voltage	$I_{TM} = 5\ A$,	See Note 6			1.7	V
dv/dt	Critical Rate of Rise of Off-State Voltage	V_D = Rated V_D,	$R_{GK} = 1\ k\Omega$, $T_C = 110°C$		10		V/μs

switching characteristics at 25°C case temperature

PARAMETER		TEST CONDITIONS	TYP	UNIT
t_{gt}	Gate-Controlled Turn-On-Time	$V_{AA} = 30\ V$, $R_L = 6\ \Omega$, $R_{GK(eff)} = 5\ k\Omega$, $V_{in} = 50\ V$, See Figure 1	1.75	μs
t_q	Circuit-Commutated Turn-Off Time	$V_{AA} = 30\ V$, $R_L = 6\ \Omega$, $I_{RM} \approx 8\ A$, See Figure 2	7.7	μs

blocked at all times. The most frequent abbreviation is I_R. The other leakage current is "anode *forward* blocking current," most often abbreviated I_F. Typical leakage current is 1 microamp. The test conditions for this specification include a given voltage. This means that, at this voltage, the device will leak no more than 1 microamp.

Breakdown voltage for an SCR is expressed as both a forward and a reverse parameter.

Breakdown voltage is the voltage at which the blocking capability fails, and a massive amount of current rushes through. SCRs use the terms "absolute maximum *forward* blocking voltage" and "absolute maximum *reverse* blocking voltage." Typically, SCRs can stand 100 volts in either direction without breaking down.

Operating speed is a measure of how quickly an SCR responds to gate pulses and interruptions in the power supply.

Operating speed of an SCR is specified in terms of "turn-on time" and "commutating turn-off time." The term "commutating" is included as a reminder that the device does not turn off by itself, but rather is turned off by an interruption of the power supply. Typical switching speeds are 1 and 2 microseconds.

In addition to these five universal specifications, the SCR has two other important parameters;

Gate trigger current and gate trigger voltage specify the current and voltage necessary to turn an SCR on.

Gate trigger current I_{GT} specifies how much *current* is required to turn the device on. In our typical example, the specification says that *no more than* 100 microamps (at the proper voltage) is required.

Gate trigger voltage V_{GT} specifies the *voltage* required to trigger the device — in our case, no more than 0.7 volts. Taken together, these last two specifications say that 100 microamps at 0.7 volt will be enough to trigger the device.

Now that we understand SCRs, the other kind of thyristor, the triac, is easy to understand. The function of the triac is identical to that of the SCR, except that triacs can be triggered into conduction in *either* direction.

HOW DOES A TRIAC DIFFER FROM AN SCR?

A triac is a five-layer semiconductor that resembles two SCRs merged together.

The triac behaves very much like a pair of SCRs connected head-to-toe. Indeed, the symbol for a triac, shown in Fig. 9.8a, reflects this. Figure 9.8b makes this clearer; the left-hand SCR in this figure can conduct conventional current in the downward direction, and the right-hand SCR can pass it in the opposite direction.

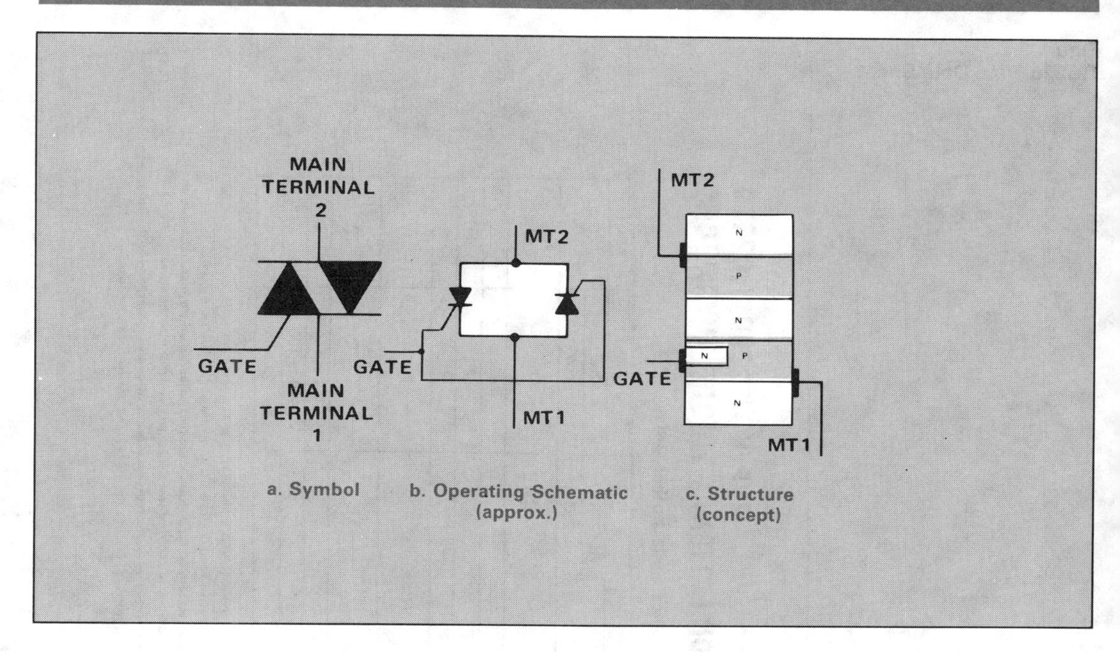

Figure 9-8.
Structure of a Triac

Like an SCR, a triac normally blocks current in both directions. Unlike an SCR, it can conduct in either direction when triggered.

Notice that the triac gate terminal is common to *both* SCRs. Picture the triac in an alternating-current circuit. It normally blocks working current attempting to pass in *either* direction between the two main terminals. But it can be *triggered* into conduction in both directions by a momentary pulse applied to the gate. This is reflected in the formal name of the triac, which is "bidirectional triode thyristor."

The triac semiconductor element has *five* layers, and a small *sixth region* under part of the gate contact, as shown in Fig. 9. 8c. This gate arrangement permits the triac to be triggered by gate current moving in *either* direction — not just *into* the gate, as with the SCR.

You'll recall that the SCR was essentially two transistors connected side by side. The triac is equivalent of two SCRs side by side. Note that the triac terminals are not called "anode" and "cathode," but rather "main terminal one" and "main terminal two." Since the operation of the triac is so similar to that of the SCR, the specifications for the two are virtually identical. Figures 9.9 and 9.10 reproduce the chief specifications from a triac data sheet; these specifications are keyed to the numbers of the SCR specifications shown in Fig. 9.5.

Figure 9-9. TIC226 Triac Data Sheet

TYPES TIC226B, TIC226D
SILICON BIDIRECTIONAL TRIODE THYRISTORS

absolute maximum ratings over operating case temperature range (unless otherwise noted)

			UNIT
Repetitive Peak Off-State Voltage, V_{DRM} (See Note 1)	TIC226B	±200	V
	TIC226D	±400	
Full-Cycle RMS On-State Current at (or below) 85°C Case Temperature, $I_{T(RMS)}$ (See Note 2)		8	A
Peak On-State Surge Current, Full-Sine-Wave, I_{TSM} (See Note 3)		±70	A
Peak On-State Surge Current, Half-Sine-Wave, I_{TSM} (See Note 4)		±80	A
Peak Gate Current, I_{GM}		±1	A
Peak Gate Power Dissipation, P_{GM}, at (or below) 85°C Case Temperature (Pulse Width ≤ 200 μs)		2.2	W
Average Gate Power Dissipation, $P_{G(av)}$, at (or below) 85°C Case Temperature (See Note 5)		0.9	W
Operating Case Temperature Range		–40 to 110	°C
Storage Temperature Range		–40 to 125	°C
Lead Temperature 1/16 Inch from Case for 10 Seconds		230	°C

NOTES:
1. These values apply bidirectionally for any value of resistance between the gate and Main Terminal 1.
2. This value applies for 50-Hz to 60-Hz full-sine-wave operation with resistive load. Above 85°C derate according to Figure 2.
3. This value applies for one 60-Hz full sine wave when the device is operating at (or below) the rated value of on-state current. Surge may be repeated after the device has returned to original thermal equilibrium. During the surge, gate control may be lost.
4. This value applies for one 60-Hz half sine wave when the device is operating at (or below) the rated value of on-state current. Surge may be repeated after the device has returned to original thermal equilibrium. During the surge, gate control may be lost.
5. This value applies for a maximum averaging time of 16.6 ms.

Figure 9-10. TIC226 Data Sheet (Continued)

electrical characteristics at 25°C case temperature (unless otherwise noted)

PARAMETER		TEST CONDITIONS	MIN	TYP	MAX	UNIT
I_{DRM}	Repetitive Peak Off-State Current	V_{DRM} = Rated V_{DRM}, $I_G = 0$, $T_C = 110°C$			±2	mA
I_{GTM}	Peak Gate Trigger Current	$V_{supply} = +12\ V$†, $R_L = 10\ \Omega$, $t_{p(g)} \geq 20\ \mu s$		15	50	mA
		$V_{supply} = +12\ V$†, $R_L = 10\ \Omega$, $t_{p(g)} \geq 20\ \mu s$		−25	−50	
		$V_{supply} = -12\ V$†, $R_L = 10\ \Omega$, $t_{p(g)} \geq 20\ \mu s$		−30	−50	
		$V_{supply} = -12\ V$†, $R_L = 10\ \Omega$, $t_{p(g)} \geq 20\ \mu s$		75		
V_{GTM}	Peak Gate Trigger Voltage	$V_{supply} = +12\ V$†, $R_L = 10\ \Omega$, $t_{p(g)} \geq 20\ \mu s$		0.9	2.5	V
		$V_{supply} = +12\ V$†, $R_L = 10\ \Omega$, $t_{p(g)} \geq 20\ \mu s$		−1.2	−2.5	
		$V_{supply} = -12\ V$†, $R_L = 10\ \Omega$, $t_{p(g)} \geq 20\ \mu s$		−1.2	−2.5	
		$V_{supply} = -12\ V$†, $R_L = 10\ \Omega$, $t_{p(g)} \geq 20\ \mu s$		1.2		
V_{TM}	Peak On-State Voltage	$I_{TM} = \pm 12\ A$, $I_G = 100\ mA$, See Note 6			±2.1	V
I_H	Holding Current	$V_{supply} = +12\ V$†, $I_G = 0$, Initiating $I_{TM} = 500\ mA$		20	60	mA
		$V_{supply} = -12\ V$†, $I_G = 0$, Initiating $I_{TM} = -500\ mA$		−30	−60	
I_L	Latching Current	$V_{supply} = +12\ V$†, See Note 7		30	70	mA
		$V_{supply} = -12\ V$†, See Note 7		−40	−70	
dv/dt	Critical Rate of Rise of Off-State Voltage	V_{DRM} = Rated V_{DRM}, $I_G = 0$, $T_C = 110°C$		500		V/μs
dv/dt	Critical Rate of Rise of Commutation Voltage	V_{DRM} = Rated V_{DRM}, $I_{TRM} = \pm 12\ A$, $T_C = 85°C$, See Figure 3	5			V/μs

†The supply voltage is called positive when it causes Main Terminal 2 to be positive with respect to Main Terminal 1.

NOTES: 6. This parameter must be measured using pulse techniques. $t_w \leq 1$ ms, duty cycle $\leq 2\%$. Voltage-sensing contacts, separate from the current-carrying contacts, are located within 0.125 inch from the device body.

7. The triacs are triggered by a 15-V (open-circuit amplitude) pulse supplied by a generator with the following characteristics: $R_G = 100\ \Omega$, $t_w = 20\ \mu s$, $t_r \leq 15$ ns, $t_f \leq 15$ ns, f = 1 kHz.

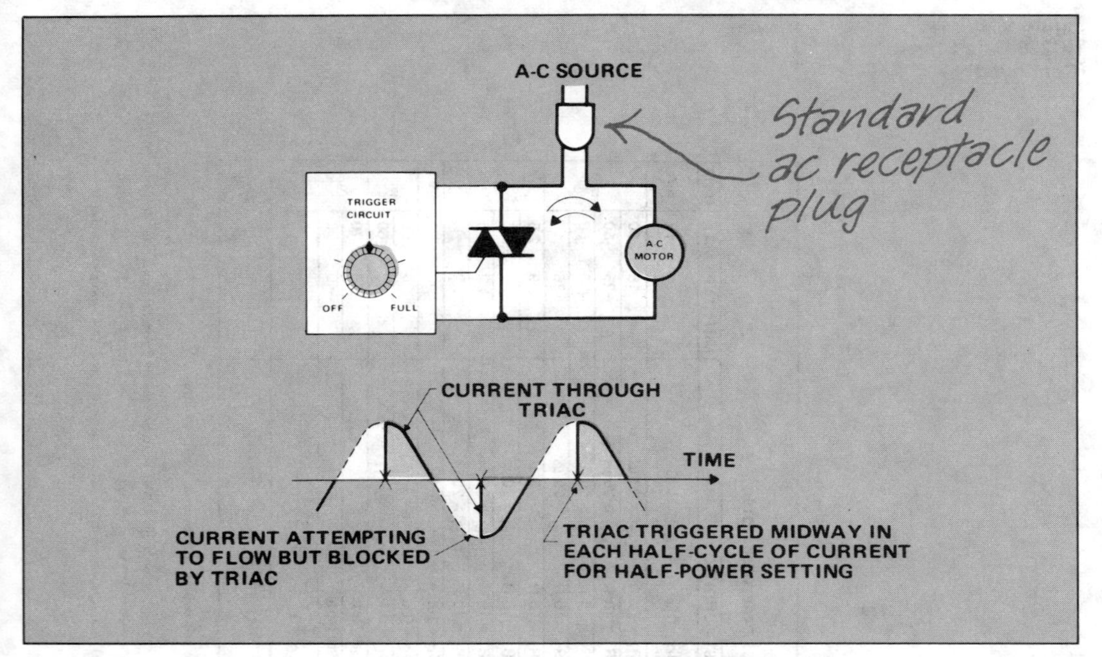

Figure 9-11. TRIAC Used for AC Motor Control

Like SCRs, triacs work well in ac phase control applications. Since triacs conduct during both positive and negative half-cycles, they are used to control ac rather than dc motors.

WHAT IS A COMMON APPLICATION OF THE TRIAC?

You'll recognize the circuit of Fig. 9.11 as being essentially the same as we used to illustrate operation of the SCR. But now, although we are still using a-c power, we are using an *a-c* motor.

The dashed curve on the waveform shows the alternating current into the triac. This is also the current entering the motor when the trigger circuit is set at *full* speed — the triac is triggered at the *beginning* of each wave, whether it is a forward or a reverse half-wave.

But if we set the knob at *half* speed, the control circuit delays triggering the triac until the *middle* of each half-wave, as shown by the solid curves. The resulting waveform is still alternating current, but since less current passes each way on the average, the motor slows down. Thus, the triac can *regulate* alternating current by means of *switching.* This is called the "phase-control" technique.

So much for thyristors. Now we take a look at some semiconductors that interface with the outside world, optoelectronic devices.

WHAT ARE OPTOELECTRONIC SEMICONDUCTORS?

Optoelectronic semiconductor devices are simply specially made diodes and transistors that interact with *light* to a useful extent. All diodes and transistors interact with light to some degree — one of the functions of their packages is to shut out light — but optoelectronic devices are designed to make *efficient use* of this phenomenon.

Optoelectronic semiconductors are designed to either sense or emit light.

There are two important categories of optoelectronic devices: *Light sensors* are diodes and transistors that convert light into electric current. *Light emitters* are diodes that convert electrical power into light (just the reverse). Figure 9.12 shows the symbols for these devices; the photodiode and phototransistor are sensors,

Figure 9-12. Optoelectronic Semiconductors

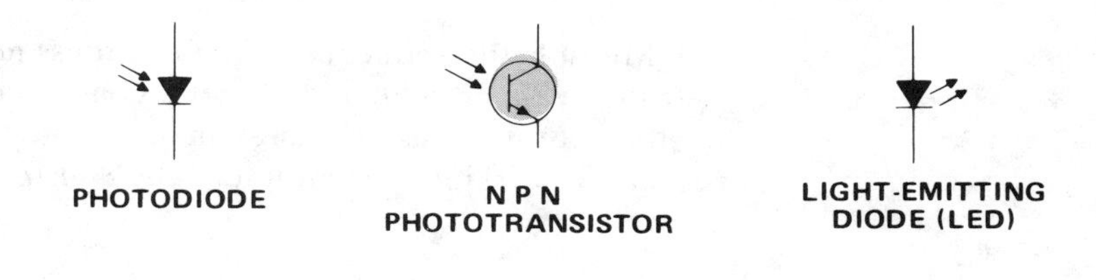

and the light-emitting diode (or LED) is a light emitter. The phototransistor symbol is especially revealing. Notice that it is just an ordinary transistor symbol without a base lead; instead, the arrows indicate light shining onto the base.

Light that strikes semiconductor material frees electrons and creates holes. Electron movement into holes creates light.

All optoelectronic devices are based on one simple principle: Whenever light strikes semiconductor material, it tends to knock bound electrons out of their sockets, so to speak, creating free electrons and holes. Conversely, when an electron falls *into* a hole, it tends to create a particle of light (a "photon").

You know what happens when free electrons and holes start roaming around the junction of a transistor or diode. Figure 9.13 shows a photodiode being used to control a very small direct-current motor. The object is to operate the motor only when light strikes the photodiode — then, the stronger the light, the faster the motor. The photodiode is much like an ordinary diode, except that it has a window or lens that lets light fall onto the PN junction. In our circuit, the battery is attempting to pump electrons from P to N through the diode. But this is the reverse direction for electrons. As you recall from our chapter on diodes, the free electrons and holes are forced apart, leaving a depletion layer around the junction, devoid of holes and free electrons. The only current that flows is leakage current, and it's negligible.

Light sensors (photodiodes and phototransistors) operate in a reverse-biased mode. Light falling on PN junctions stimulates the flow of leakage current.

Now suppose a photon of light shoots into the semiconductor element, within the area of the depletion layer. Then a free electron and a hole are created in the depletion layer. (This assumes that the diode is properly designed, uses the right kind of semiconductor material — usually silicon — has enough of the right dopants, and has a chip shaped and placed properly to receive the light, etc.) Immediately, the free electron is forced by the battery into the N region and out the cathode terminal, while the hole is driven in the opposite direction. The net result is that one electron passes from P to N, through the circuit and the motor.

Multiply this sequence by the countless millions of photons in a strong beam of light, and we get a considerable current to drive the tiny motor. Thus, the function of a photodiode is to switch and regulate a working current under the *control* of light striking the device.

The effect to remember is that light shining on a semiconductor junction greatly increases the reverse *leakage* current. This applies to all semiconductor devices. Light sensors are merely diodes and transistors in which this effect is enhanced and put to efficient use.

In the case of photodiodes, there is one qualification to the rule that light increases the leakage current. It is true that light falling on a PN junction always tends to make electrons flow in the reverse direction, from P to N. But the fact is that an external power supply is not always required to make current flow. In the circuit of Fig. 9.13, if we remove the battery, so that the circuit

Figure 9-13. Photodiode Used for DC Motor Control

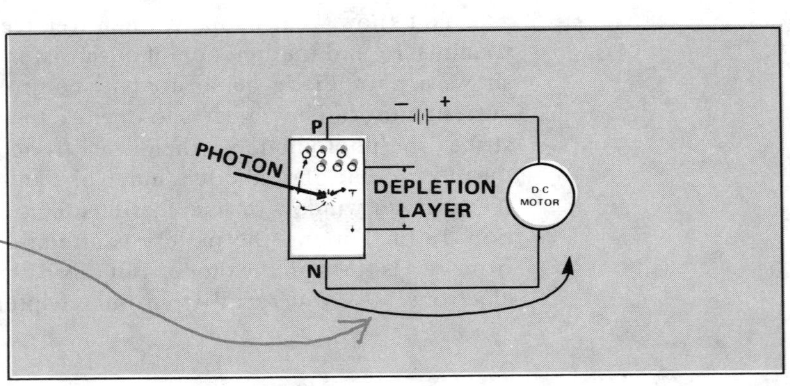

One electron flows to motor for every proton that strikes.

consists only of the photodiode and the motor, then a small amount of current will flow when light strikes the photodiode — generated entirely by light.

Photodiodes do not always need external power supplies to ensure current flow. "Photoconductive" diodes are used with separate power sources, but "photovoltaic" diodes generate current by themselves.

To put it simply, every photodiode is theoretically capable of converting light energy into electrical energy. Solar cells in artificial satellites and the sensors in many light meters are simply photodiodes especially constructed to enhance this capability of generating electric current. When photodiodes are used along with a separate power supply (as in Fig. 9.13), they are called "photoconductive," meaning they *conduct* current when illuminated, and block current when dark. But photodiodes used to *generate* current without the assistance of another power supply are called "photovoltaic" — because they actually produce a voltage pressure in the reverse direction.

WHAT DO PHOTOTRANSISTORS DO THAT PHOTODIODES DON'T?

A phototransistor operates like a photoconductive diode, but the small leakage current induced by light is amplified into a large working current.

The phototransistor, like most photodiodes, is a "photoconductive" device. Figure 9.14 shows an NPN (a PNP version is possible too, of course). We could replace the photodiode in the circuit of Fig. 9.13 with this phototransistor. It, too, functions to switch and regulate current as the power supply attempts to pump electrons from emitter to collector. When there is no light, no current flows, because there is no base *control* current. But when light strikes the base-collector junction, then reverse leakage current flows. This, in effect, constitutes a current of electrons

Figure 9-14. NPN Phototransistor Cross-Section

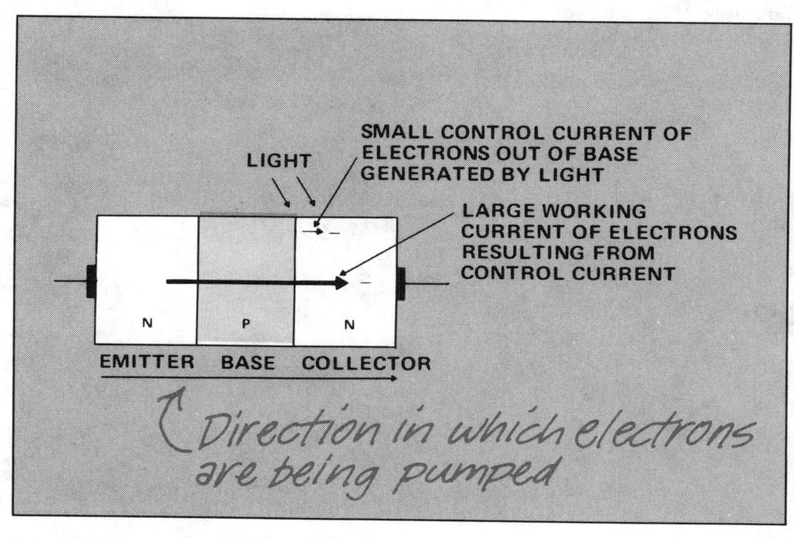

being withdrawn from the base — just what we need to turn on an NPN transistor. But in the phototransistor, unlike the diode, a much *larger* working current flows from emitter to collector. Thus, the phototransistor works like the photodiode, but in addition *amplifies* the tiny current produced by the light. Recall that with the diode, we get only *one* electron of current, for each photon of light. With the phototransistor, however, each electron leaving the P region allows perhaps a *hundred* electrons to pass from emitter to collector and through the working circuit.

HOW DO LIGHT EMITTERS WORK?

Electrons falling into holes create light. Light-emitting diodes (LEDs) are designed to maximize this effect.

Light-emitting diodes simply reverse the effect utilized in light sensors. When a free electron in a piece of semiconductor material meets a hole and falls into it, the process generates a photon of light. The photon is hurled away in some random direction. Countless photons escaping together constitute a ray of light.

Figure 9.15 shows how this effect is used in light-emitting diodes to produce a useful amount of light, so that an LED serves as a sort of semiconductor light bulb. The LED is connected into a circuit with a power supply pumping forward current through it. Electrons injected into the chip at the cathode cross the N region as free electrons, while holes created at the anode by the withdrawal of bound electrons cross the P region. Near the PN junc-

Figure 9-15. Structure and Operation of a Light-Emitting Diode

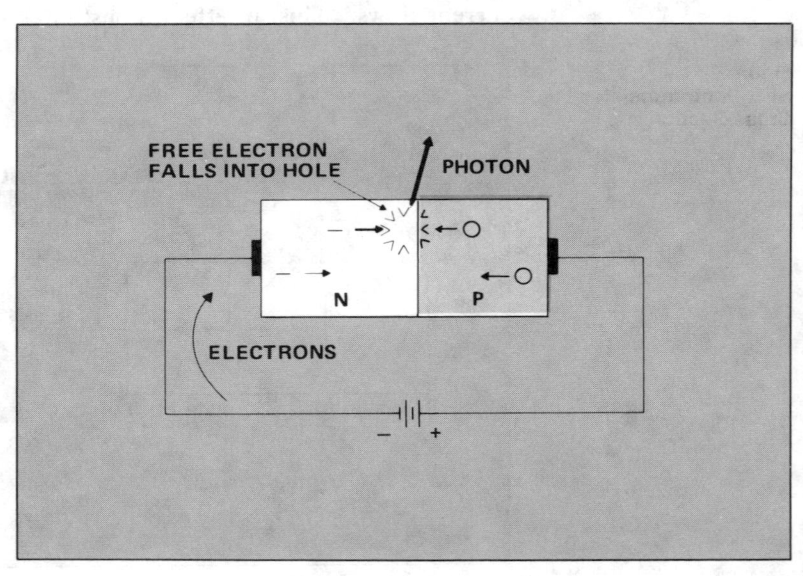

tion, free electrons fall into holes, generating photons of light in the process. Given enough forward current, the junction area of the chip glows brightly.

LEDs are manufactured from special compounds such as gallium arsenide, which make better light emitters than do silicon or germanium.

Light-emitting diodes are not made of silicon or germanium. These two substances, while quite suitable for light-*sensing*, are too inefficient to make good light-*emitters*. They emit more heat than light. So instead, a compound of elements is used as the semiconductor substance. The most popular one is gallium arsenide. Gallium is a P-type dopant, and arsenic is an N-type dopant. Combined in precisely equal quantities, they exactly cancel out each other's dopant effect and provide a structure much like pure silicon. Gallium arsenide in its *pure* form has very few positive holes or free electrons. But just as is the case using silicon, the gallium arsenide must be *doped* to produce P-type material in one area of the chip, and N-type material in the other.

WHAT COLORS OF LIGHT ARE INVOLVED IN OPTOELECTRONICS?

Depending on the intended application, optoelectronic devices work with either infra-red or visible light. Light sensors are most sensitive to infra-red light.

Typical LEDs produce *infra-red* light, a range of color that is invisible to the human eye. But as it turns out, most semiconductor light-*sensors* are most sensitive to just this color range of light. The result is that these infra-red devices lend themselves to such interesting and useful applications as burglar alarms and military night-time surveillance systems.

Many light sensors and emitters are available, however, that sense or generate *visible* light — usually red light. Visible light emitting diodes (VLEDs) are already widely used to display digits in desk-top calculators and other instruments. You'll recall that we used them as the display elements in our baby computer.

HOW ELSE ARE OPTOELECTRONIC DEVICES USED?

Perhaps the most common use for optoelectronic light emitters and sensors is to read punched cards and punched paper tape. The familiar Hollerith card, of the kind used commonly to bill retail charge customers, is still a very popular medium for inputting data to computer systems.

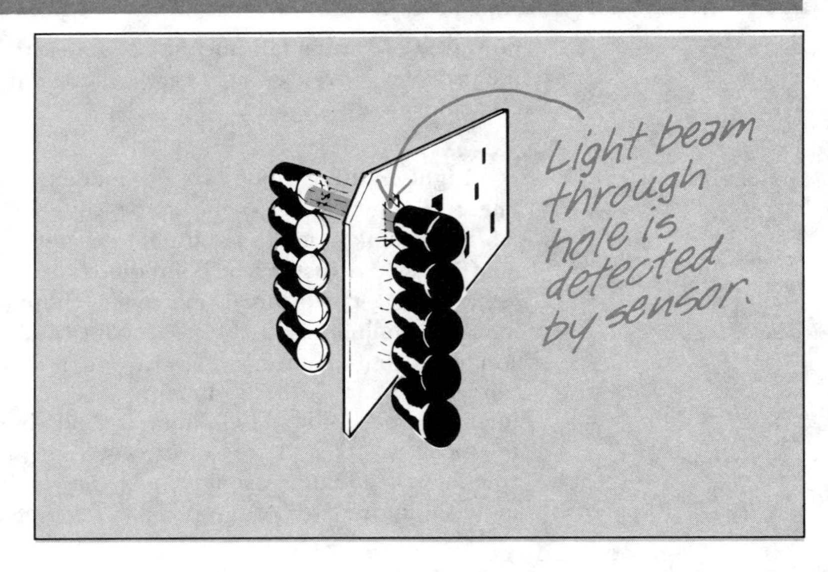

Figure 9-16. Operation of an Optoelectronic Card Reader

Among the many uses for light emitters and sensors are card and tape readers, burglar alarms, and devices for industrial inspection or measurement.

Figure 9.16 suggests how an optoelectronic card reader works. Each hole in the punched card represents one bit of data (one "datum"). The card is passed between an array of light emitters on one side and light sensors on the other. Each emitter generates light continuously, but each sensor is actuated only when a hole passes by and admits light to it. In this way, each hole transmits light to the sensor and creates a pulse of electricity; this pulse is properly interpreted by the computer and used as input data. This optoelectronics technique is much more reliable than the old-fashioned contact method, because with only light beams touching the card, there are no moving parts to wear out or stick.

This concept of interrupting a beam of light between an emitter and a sensor is also the basis, of course, for many burglar and intruder detectors. In such applications, the ability of solid-state optoelectronics to work so well with infra-red light is put to good use. In such systems, the intruder is unaware that his presence has been detected.

As process improvements bring down the prices of optoelectronic devices, they are finding more and more uses in industry — to count objects on a conveyer belt as the objects interrupt the light beam, in inspection, in measuring, in flow control, weight control, level indicators, and many more applications.

WHAT SPECIFICATIONS ARE FOUND ON OPTOELECTRONIC DATA SHEETS?

Many of the specifications applied to optoelectronic devices are the same as the ones you're already familiar with from our discussions of ordinary diodes and transistors — so there's no need to repeat them. But as you would expect, optoelectronic devices require some additional specifications dealing with light.

For **light sensors:**

Light current I_L is the amount of current that flows (under certain voltage conditions) under conditions of some stated amount of *light.*

Dark current I_D is the amount of current that flows (under certain voltage conditions) under the condition of *darkness.*

For **light emitters:**

Radiant power output P_O is the amount of light power (the brightness) that the diode produces, given a certain forward-current test condition. A fairly typical specification is one milliwatt of radiant power.

Wavelength at peak emission λ_{peak} is the color of the brightest color emitted. The upside-down "y" is the Greek letter "lambda," or "L." Here it stands for *length,* meaning wavelength. So this specification tells us the wavelength of the brightest color from the light emitter. A typical wavelength is 0.93 micrometer.

Spectral bandwidth BW_{λ} is a measure of how sharp or concentrated the color is. Since the light source emits other colors on either side of its peak color, this number tells us how broad a portion of the spectrum is emitted, including all colors emitted at least half as brightly as the peak color. A typical value is 500 Ångstroms.

Figure 9.17 is an excerpt from a light-emitter data sheet, and shows the most common specifications. Figure 9.18 is an excerpt from a photodiode data sheet, and Figs. 9.19 and 9.20 are part of a phototransistor data sheet. Studying these data sheets and relating them to our discussions of semiconductor specifications will go a long way toward helping you to understand semiconductors.

This is our last chapter on single semiconductors in individual packages — called "discrete" devices. In the next chapter, we'll start talking about *integrated* circuits, devices that contain many

Figure 9-17. Light Emitter Data Sheet

TYPES TIL23, TIL24
P-N GALLIUM ARSENIDE LIGHT SOURCES

absolute maximum ratings

Reverse Voltage at 25°C Case Temperature . . . . 2 V
Continuous Forward Current at 25°C Case Temperature (See Note 1) . . . . 100 mA
Operating Case Temperature Range . . . . −65°C to 125°C
Storage Temperature Range . . . . −65°C to 150°C
Soldering Temperature (3 Minutes) . . . . 240°C

NOTE 1: Derate linearly to 125°C case temperature at the rate of 1 mA/°C. For pulsed operation at higher currents, see Figures 8 and 9.

operating characteristics at 25°C case temperature

	PARAMETER	TEST CONDITIONS	TIL23 MIN	TIL23 TYP	TIL23 MAX	TIL24 MIN	TIL24 TYP	TIL24 MAX	UNIT
P_O	Radiant Power Output	$I_F = 50$ mA	0.4			1			mW
λ_{peak}	Wavelength at Peak Emission			0.93			0.93		μm
BW_λ	Spectral Bandwidth between Half-Power Points			500			500		Å
θ_{HP}	Emission Beam Angle between Half-Power Points			35°			35°		
V_F	Static Forward Voltage			1.25	1.5			1.5	V

diode and transistor elements, along with resistors and connectors, all connected together into a functional circuit, entirely within a single chip of silicon. Now you see where all of our discussion has been leading — and when we discuss *integrated circuitry,* you will see where the entire evolution of semiconductor electronics has led us.

Figure 9-18.
Photodiode Data Sheet

TYPE TIL77
P-N SILICON PHOTODIODE

absolute maximum ratings at 25°C free-air temperature (unless otherwise noted)

Forward Voltage . . . . 0.4 V
Reverse Voltage . . . . 10 V
Operating Free-Air Temperature Range . . . . . . −20°C to 80°C
Storage Temperature Range . . . . . . −20°C to 80°C
Lead Temperature 1/16 Inch from Case for 10 Seconds . . . . . . 240°C

electrical characteristics at 25°C free-air temperature (unless otherwise noted)

	PARAMETER		TEST CONDITIONS	MIN	TYP	MAX	UNIT
$V_{(BR)}$	Breakdown Voltage		$I_R = 10\ \mu A$, $E_v = 0$	10			V
I_D	Dark Current		$V_R = 3\ V$, $E_v = 0$		0.25	5	nA
I_L	Light Current	Photoconductive Operation	$V_R = 3\ V$, $E_v = 50\ lm/ft^2$†	1.5	2		µA
		Photovoltaic Operation	$V_R = 0$, $E_v = 50\ lm/ft^2$†		2		
V_F	Forward Voltage		$I_F = 10\ \mu A$, $E_v = 0$	0.4			V
λ_{peak}	Wavelength at Peak Response				0.57		µm
BW_λ	Spectral Bandwidth between 20-Percent Response Points		$V_R = 3\ V$		0.3		µm
C_T	Total Capacitance		$V_R = 3\ V$, $E_v = 0$, $f = 1\ MHz$		750		pF

†The common British-American unit of illumination is the lumen per square foot. The name footcandle has been used for this unit in the USA. To convert to units of the International System (SI), use the equivalency:

1 lumen per square foot = 10.764 lux (lumens per square meter).

Figure 9-19. Phototransistor Data Sheet

TYPES TIL601 THRU TIL616
N-P-N PLANAR SILICON PHOTOTRANSISTORS

absolute maximum ratings at 25°C case temperature (unless otherwise noted)

Collector-Emitter Voltage . . . . 50 V
Emitter-Collector Voltage . . . . 7 V
Continuous Device Dissipation at (or below) 25°C Case Temperature (See Note 1) . . . . 50 mW
Operating Case Temperature Range . . . . −65°C to 125°C
Storage Temperature Range . . . . −65°C to 150°C
Soldering Temperature (3 minutes) . . . . 240°C

electrical characteristics at 25°C case temperature (unless otherwise noted)

PARAMETER		TEST CONDITIONS	TYPE	MIN	TYP	MAX	UNIT
$V_{(BR)CEO}$	Collector-Emitter Breakdown Voltage	$I_C = 100\ \mu A$, $H = 0$	ALL	50			V
$V_{(BR)ECO}$	Emitter-Collector Breakdown Voltage	$I_E = 100\ \mu A$, $H = 0$	ALL	7			V
I_L	Light Current	$V_{CE} = 5$ V, $H = 20$ mW/cm^2, See Note 2	TIL601 TIL605 TIL609 TIL613	0.5		3	mA
			TIL602 TIL606 TIL610 TIL614	2		5	mA

Figure 9-20. Phototransistor Data Sheet (Continued)

			TIL603 TIL607 TIL611 TIL615	4		8	mA
			TIL604 TIL608 TIL612 TIL616	7			mA
I_D	Dark Current	V_{CE} = 30 V, H = 0	ALL			25	nA
		V_{CE} = 30 V, H = 0, T_C = 100°C	ALL		1		μA
$V_{CE(sat)}$	Collector-Emitter Saturation Voltage	I_C = 0.4 mA, H = 20 mW/cm², See Note 2	ALL		0.15		V

NOTES: 1. Derate linearly to 125°C at the rate of 0.5 mW/°C.
2. Irradiance (H) is the radiant power per unit area incident upon a surface. For this measurement the source is an unfiltered tungsten linear-filament lamp operating at a color temperature of 2870°K.

switching characteristics at 25° C case temperature

PARAMETER		TEST CONDITIONS	TYP	UNIT
t_r	Rise Time	V_{CC} = 30 V, I_L = 800 μA, R_L = 1 kΩ, See Figure 1	1.5	μs
t_f	Fall Time		15	

Quiz for Chapter 9

1. The distinguishing feature and principal advantage of thyristors is that:

- ☐ a. They are the only high-power semiconductor switching devices available
- ☐ b. Their switching speed is faster than that of most transistors
- ☐ c. They are typically less expensive than transistors
- ☐ d. They are easy to turn off by means of control current at the gate
- ☐ e. Only a quick, momentary pulse of gate current is required to turn them on all the way (so that less power is wasted in the control circuit than with a transistor)

2. Which of the schematic symbols presented in this lesson (shown below) is that of the SCR (silicon controlled rectifier, also called "reverse-blocking triode thyristor")?

☐ a.  

☐ b. 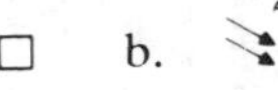

☐ c. 

☐ d. 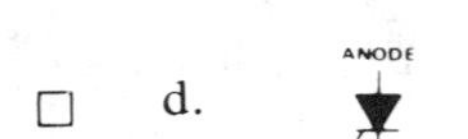

☐ e. 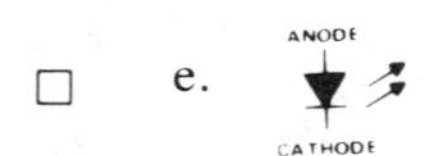

3. What is the main difference between the function of the SCR and that of the triac?

- ☐ a. The SCR performs better than the triac
- ☐ b. The SCR is a switching device, while the triac is an amplifying device
- ☐ c. The SCR can be triggered into conduction in only one direction, while the triac can be switched on for current in either direction
- d. The SCR is made of silicon but the triac is not
- ☐ e. Inside every triac package are two little SCR packages

4. What is the main advantage of the phase-control method of regulating power from an alternating-current source (in which a switching device — by precise, repetitive operation — cuts off a variable portion of each wave or half-wave)?

- ☐ a. Very little power is dissipated (wasted) by conversion into heat, because the working current is never throttled partially — a process which always generates heat
- ☐ b. It is equally applicable to ac and dc power supplies
- ☐ c. It produces smooth, constant output current to the working device
- ☐ d. It always rectifies alternating current in addition to regulating it
- ☐ e. The SCR or triac can be used alone — without any accessory trigger circuit

5. The data-sheet specifications of SCRs and those for triacs:

- ☐ a. Bear no resemblance to those of any other semiconductor devices
- ☐ b. Contain versions of the five "universal" semiconductor specifications plus others — notably, specifications on the gate current and voltage required to trigger the devices
- ☐ c. Are very different from each other
- ☐ d. Include noise figure (NF) and current gain (h_{FE}).
- ☐ e. Contain specifications for leakage current passing in only one direction

6. Photodiodes and phototransistors are used:

- ☐ a. To sense light by permitting current from some power source to pass in rough proportion to the intensity of light falling on the device
- ☐ b. To generate light for transmitting information or providing useful illumination
- ☐ c. To block current when light strikes them, and pass current when dark
- ☐ d. In the "act" section of systems
- ☐ e. For the same purposes as standard diodes and transistors

7. Which device uses the principle that light striking a PN junction knocks bound electrons out of their "sockets" and thus greatly increases the reverse leakage current?

- ☐ a. The photodiode
- ☐ b. The phototransistor
- ☐ c. The light-emitting diode (LED)
- ☐ d. All of the above
- ☐ e. a and b above

8. Light in an LED is generated by:

- ☐ a. Intense heating of the junction, like the filament of a light bulb
- ☐ b. Bound electrons spontaneously leaping out of their "sockets," creating a photon of light
- ☐ c. Free electrons falling into holes and giving up their energy in the form of a particle of light called a photon
- ☐ d. The creation of a free electron and a hole
- ☐ e. The conversion of invisible infared light from the surroundings, into visible light

9. Optoelectronic devices are particularly useful in nighttime surveillance systems and burglar alarms because

- ☐ a. The light generated by most LEDs is of an invisible infrared color
- ☐ b. Typical light sensors are most sensitive to invisible infrared color
- ☐ c. The devices are very inexpensive
- ☐ d. The existence of these devices is not very widely known
- ☐ e. a and b above

10. The data-sheet specifications of optoelectronic devices include

- ☐ a. Several specifications that apply to ordinary diodes and transistors
- ☐ b. Light current I_L and dark current I_D for sensors
- ☐ c. Wavelength specifications describing the color of light emitted by LEDs
- ☐ d. Current gain (beta, or h_{FA}).
- ☐ e. All but d above

10 INTRODUCTION TO INTEGRATED CIRCUITS

Key Words

Digital Integrated Circuit
Integrated Circuit (IC)
Industrial-Grade IC
Linear Integrated Circuit
Military-Grade IC
Photomask
Photoresist

Definitions are found in the glossary in the back of the book.

Introduction to Integrated Circuits

Integrated circuits are so important that we've devoted the rest of this book to them. Since they use diodes and transistors that function just like the ones you've already studied, there are no new specifications to learn. Rather, we need to see how these devices are put together on one piece of silicon to perform a complete circuit function.

WHAT IS AN INTEGRATED CIRCUIT?

An integrated circuit is a complete electronic circuit on a single semiconductor chip.

An integrated circuit (or IC) is a complete electronic circuit, containing transistors and perhaps diodes, resistors, and capacitors, along with their interconnecting electrical conductors, processed on and contained entirely within a single chip of silicon. The integrated circuit was invented at Texas Instruments in 1958 by Jack Kilby. Figure 10.1 shows the general appearance of a typical integrated circuit. This phantom view shows how it would look if the plastic package were semi-transparent. The working heart of the device is the tiny chip of silicon at the center. It's only a little larger than the chip used for a typical small-signal discrete transistor. IC chips come in larger sizes as we will see later on, but the size of the chip in this figure and the next are not unusual. They measure roughly 0.060 inch — about 1-16 inch square.

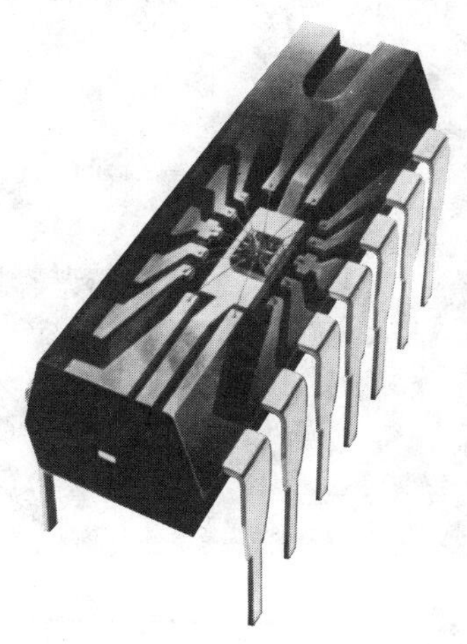

Figure 10-1. Packaged Integrated Circuit

All the additional bulk of the device simply provides communication with, and protection against, the world outside. The heavy metal leads are stiff enough for insertion into printed circuit cards, and the rugged plastic package can actually be hammered on without damaging the integrated circuit.

A typical chip is shown greatly magnified in Fig. 10.2. This is a very simple chip, used for illustration — most IC chips are bigger and contain far more components than this one. The labels on the figure point out some of the more important structures.

An integrated circuit contains the elements of a conventional discrete circuit in miniature form.

The electrical circuitry within an IC is a miniature version of a circuit that could just as well be constructed with ordinary discrete components. In fact, in the early days of IC development, circuits were first built in conventional form using discrete components — that is, they were "breadboarded," to make sure they would function properly, before being rendered in IC form. Every junction and every connection has its counterpart in the IC.

But as we have pointed out in discussing circuits and systems, it is enough for most purposes to think of circuits — and this includes integrated circuits — in terms of *building blocks*. In other words, we need to think of what the circuits do and how well they perform, without being concerned about how their components work together. The IC building block represented in Fig. 10.2, for example, is simply three NAND gates having three inputs each —

**Figure 10-2.
Integrated Circuit Chip**

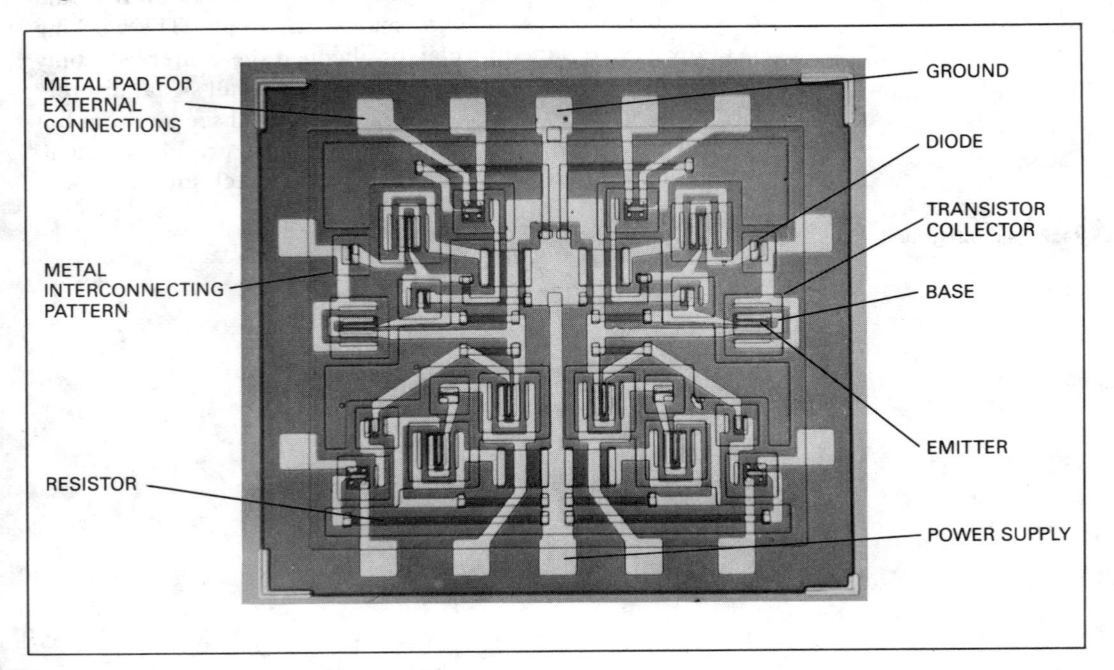

and having certain performance specifications listed in its data sheet.

In IC form, switching and amplifying circuits respectively are called digital and linear ICs.

Like all electrical circuits, integrated circuits are categorized into two classes: *Switching* and *amplifying*. In the IC world, switching circuits are customarily referred to as *digital*, and amplifying circuits are referred to as *linear*.

WHAT ARE THE ADVANTAGES OF INTEGRATED CIRCUITS?

Integrated circuits offer the advantages of small size, low cost, and high reliability.

There are three major advantages in making an electronic circuit in integrated rather than discrete form: small *size*, low *cost*, and high *reliability*.

The advantages of small size go far beyond the obvious cases of such miniature systems as hearing aids and spacecraft. Miniature circuitry permits very complicated systems such as large computers to have manageable physical dimensions — filling a room instead of a warehouse. Not only does this save money on system hardware, cabinets, wire, and floor space — but small circuits consume less power, so they require a much smaller investment in power supply and air-conditioning equipment. Furthermore, the operating speed of a system is increased by reducing its size, because the travel time for information between parts of the system is shorter. As a result of this, a smaller system can perform more decision-making tasks in a given amount of time than a larger version of the same system.

The cost savings resulting from the use of integrated circuits are not only those that are made possible by the smaller size, however. A large part of the economy comes from the decreased manufacturing cost of the circuits themselves. The cost of processing a semiconductor chip — whether it holds a single transistor or a complex IC — is roughly proportional to its area, because approximately the same number of manufacturing steps are involved in producing a slice. So by packing more components into less chip area, the cost per component is reduced. And adding even more savings is the fact that ICs mean fewer parts to order, to inventory, and to assemble into the system.

Yet as important as the advantages of small size and low cost are, they are overshadowed by the advantage of greatly improved reliability. Reliability means simply that a circuit or a system will perform for a long time without suffering poor performance or breakdowns. IC systems are much less likely to fail than are discrete versions of the same system. The major reason for the extremely high reliability of IC systems is that they require far fewer solder joints and mechanical connections. In any semiconductor system, it is the failure of these interconnections — *not* the failure of the components — that accounts for most failures. Moreover, since the use of ICs means fewer separately assembled components, there is less likelihood of mistakes in assembly, or of bad devices being used.

In a larger sense, the advantages of integrated circuits are best envisioned in terms of the historical trend of systems toward ever greater complexity. More and more components and circuit functions are being used per system. Our civilization demands more complex electronic systems year after year. In order to achieve this complexity, the *cost* and *size* of the systems must be kept within reasonable bounds. And the *reliability* of each part must continually improve so that the systems will function without breaking down every few minutes. Integrated circuits offer giant steps toward meeting this challenge of complexity.

WHAT ARE THE LIMITATIONS OF INTEGRATED CIRCUITS?

Due to their small size, ICs have limitations in power, voltage, and the types of components they can contain.

Despite the amazing advantages of ICs, and capabilities that continue to broaden each year and displace more and more discrete electronic components, ICs still have definite limitations (at this writing). Most important, ICs have been limited to comparatively low power, low voltage, and a limited selection of components that can be integrated economically.

The *power* limitation results mainly from the small size of ICs. The more current a device carries, the more heat it generates, as we have seen. If this heat is concentrated in a tiny device, it produces temperatures high enough to damage or destroy the element. Consequently, the size advantage of ICs is a trade-off that sacrifices current-handling capability. This limits most ICs to applications involving *information* rather than *work*. Even modern computers that use ICs for handling information still use discrete high-power transistor circuits for output to working devices.

By the same token, *voltages* in ICs must be kept fairly low, because the insulation between circuit elements is relatively weak, owing to the components being packed very close together in one chip of material. Typical voltage rating of an IC is between five and twenty volts. If this voltage is exceeded, the insulation breaks down in one place or another, resulting in a short-circuit.

Certain components, such as resistors and capacitors, are difficult to fabricate on IC chips. Integrated circuits use as few of such components as possible.

We say the selection of components is limited because, although the silicon of an IC chip is ideal material for transistors and diodes, it doesn't work too well for other circuit components. Resistors, for example, tend to turn out with the wrong value of resistance. Despite the most precise controls, resistance tolerances of 25 % are typical — compared to tolerances of 1 % for inexpensive discrete resistors. Moreover, the higher the resistance value, the more space a resistor takes in an integrated circuit; a 40,000-ohm resistor represents the approximate economical limit, and takes up as much space as several transistors. Capacitors pose even more of a space problem than resistors. These temporary storage reservoirs for electrons demand large areas. As small a capacitance value as 20 picofarads (20 trillionths of a farad) demands more space than several transistors. Unfortunately, many common electronic circuits require capacitances millions of times as great. Similarly, inductors such as transformers are almost impossible to produce on a silicon chip.

Because of the limitations, typical electronic systems use as few of these hard-to-integrate components as possible. But they still require some discrete resistors, capacitors, inductors, transformers, and high-power semiconductor devices. Ingenious systems designers have worked around these limitations by creating circuits that use low-power transistors and diodes — which are easy and cheap to integrate — to build circuits that *function* just as well as those that demand difficult-to-integrate components.

The rewards are great, and systems in which the difficult components have been eliminated most completely are the types of systems that are most thoroughly integrated — powerful digital computers are the classic example of this trend. And working from their end, the manufacturers of integrated circuits are continually advancing the application boundaries, especially in the area of power-handling capability.

And despite these limitations, the dollar value of ICs sold is already roughly equal to that of discrete devices at this writing — ICs are the wave of the future.

HOW ARE INTEGRATED CIRCUIT CHIPS MANUFACTURED?

We're on familiar ground again, because IC chips are made by the planar diffusion techniques we discussed earlier. A typical IC

Using planar diffusion techniques, IC manufacturers produce hundreds of integrated circuits at a time from each slice of silicon.

slice two inches in diameter (Fig. 10.3), when completely diffused, may contain hundreds of individual IC chips, each a complete circuit in itself. The integrated circuit slice here contains more than 600 integrated circuits, known as type "SN7400." Each of the circuits is about as big as the "M" on the dime. (The incomplete circuits around the rough edge of the slice are useless, of course.) When the diameter of the slice is increased to 3 inches, the number of circuits per slice is more than doubled. More typically, the number of circuits per slice remains at around 600 while the complexity of each circuit is more than doubled. Similarly, increasing the slice diameter to 4 inches allows the fabrication of the same number of even more complex devices. Using a 4 inch slice rather than a 2 inch slice alternatively, four times as many devices of a given complexity can be fabricated.

The concept of the structure of a simple circuit on an integrated circuit chip is shown in Fig. 10.4. This is not a particularly useful circuit, but it serves as illustration. It demonstrates the structure and interconnection technique for the three most common IC components — a resistor, a diode, and an NPN transistor.

Applications of light and chemicals create "windows" through silicon oxide to the silicon substrate, through which successive diffusions occur. Finally, a pattern of metal leads is etched over the chip, making it possible to attach wire leads to the device.

We start with a silicon slice cut from a crystal grown as P type. Then three successive diffusions of dopants (N,P,N) are made through appropriately shaped windows cut in the silicon oxide film by acid. After each diffusion, the windows are covered by growing a new oxide layer. After a final set of windows is cut, a layer of gold or aluminum metallization is deposited over the oxide, penetrating through the last set of windows to contact the silicon.

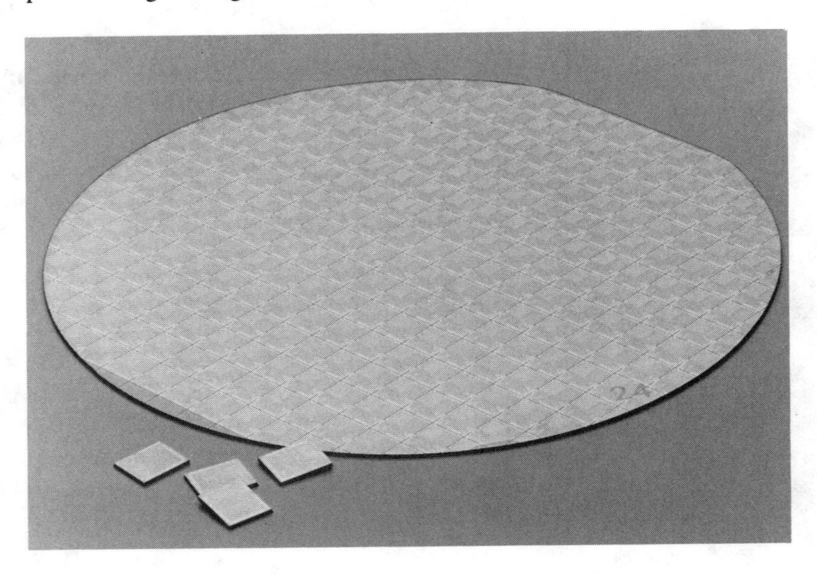

Figure 10-3. A Complete Slice of Integrated Circuits

Figure 10-4.
Integrated Circuit Components Formed by Multiple Diffusions

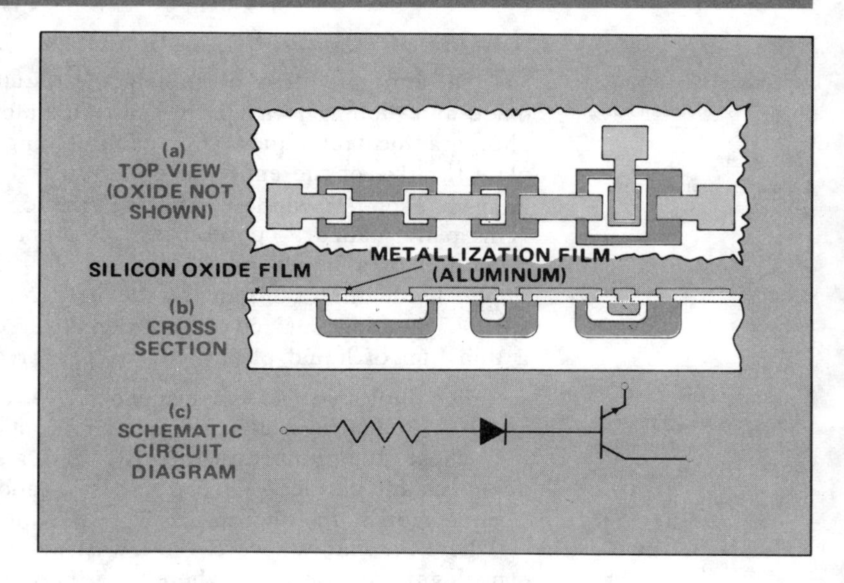

Acid is once again used to etch away all but certain desired strips of the metal. This forms a pattern of electrical leads, and also forms large "bonding-pad" areas at the ends of some leads, where fine gold wires are attached during the assembly process, after the chip has been cut from the slice.

The structure of the NPN transistor and the PN diode are obvious in Fig. 10.4. The resistor is simply a long, narrow strip of P-type material (surrounded by an N region) through which current is channeled by means of metallized connections at both ends. The desired value of resistance can be achieved by adjusting the length and width of the P-type resistor diffusion during the design of the circuit.

Since all the components are part of the same silicon slice, it may appear that they would be short-circuited together. The cross-section view, however, shows that each is electrically isolated (insulated) from the other by the one-way-valve action of PN junctions. For conventional current to take the undesired short-circuit path, for example, from the P-type resistor to the N-type transistor collector region, it would have to pass from P to N to P to N. However, the one N-to-P crossing in this route is forbidden by diode action — so this short-circuit current is blocked. All the other possible short-circuit paths are similarly blocked by PN junctions. As a result, current is forced to take the intended path through the components.

HOW IS PHOTOMASKING ACCOMPLISHED?

A photographic technique known as photomasking transfers the image of the desired circuit pattern from a transparent plate to the silicon slice.

The intricate steps of etching the required windows in the oxide and etching away parts of the metallization are done by photographic techniques. For each etching step, a transparent plate the size of the entire slice is prepared. On this plate, the image of each tiny window appears in the form of an opaque spot. (This plate, called a "photomask," is prepared by photographic reduction of a larger piece of artwork; the reduction is then duplicated hundreds of times on the plate, once for each IC on the slice.) The upper surface of the oxide-covered slice is coated with a thin film of liquid plastic called "photoresist."

The silicon slice is first treated with a film of light-sensitive "photoresist." Next, ultraviolet light is shone onto the slice through the photomask. The portions of photoresist that remain in shadow are then washed away, exposing the areas of the slice where diffusion will occur.

The photomask is then suspended over the coated slice as in Fig. 10.5, and ultraviolet light is cast on the slice from above. This light causes the photoresist to harden into a solid layer of tough, acid-resistant plastic — *except* where shadows are cast by the opaque spots in the photomask. The spots of non-resistant plastic — where the shadows were — are washed away in a solvent bath, exposing the silicon oxide where the windows are to be. Then the slice is dipped in acid, and this etches away the oxide film except where it is protected by hardened photoresist. Now we have the desired windows in the oxide. The solid photoresist is stripped away with a special solvent, and the slice is ready for the diffusion furnace. This photomasking technique is the key step in the planar diffusion process that is at the heart of the semiconductor industry.

Figure 10-5. Photomasking Process

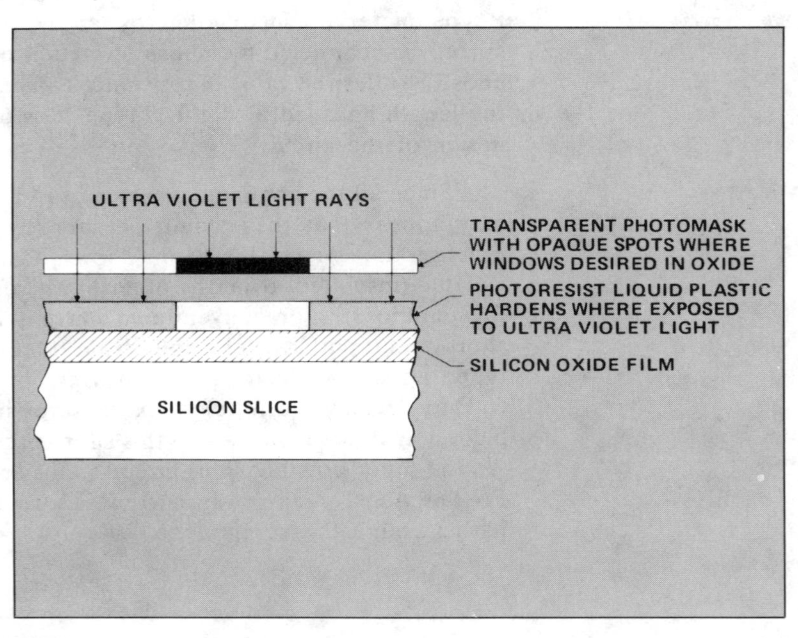

HOW ARE INTEGRATED CIRCUITS PACKAGED?

When the diffusion process is complete, the chips are mounted, wire-bonded, and packaged in metal, ceramic, or plastic.

IC chips are mounted and wire-bonded in much the same way as discrete transistors are — see Fig. 10.1. Many different package types are available; each type has its own set of advantages and limitations. Figure 10.6 presents a representative sampling of package types. The most popular at this writing are the several sizes of plastic packages ("dual in-line" refers to the two lines of leads in these package types).

One of the reasons for these packages of different size is that more complex integrated circuits, although the chip they require may not be much larger, require more leads and hence a larger package. This is a good point at which to explain three terms you may have run across in your reading:

Small-scale integration (SSI) refers to ICs having up to 9 gates.

Figure 10-6.
Types of Integrated Circuit Packages

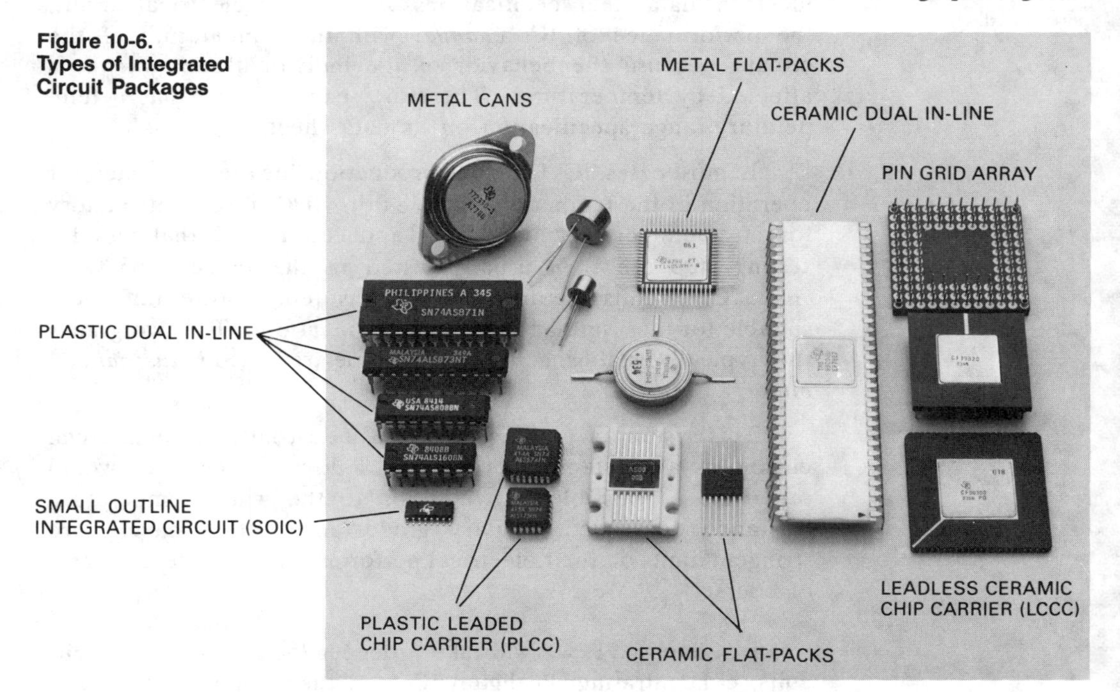

Medium-scale integration (MSI) refers to integrated circuits having 10 to 100 gates.

Large-scale integration (LSI) refers to integrated circuits having more than 100 gates.

Very large-scale integration (VLSI) refers to integrated Circuits having more than 1,000 gates.

As we will see later, VLSI devices are pushing the upper limits of complexity even higher, and working circuits containing fifty thousand gates are already commonplace.

HOW ARE INTEGRATED CIRCUITS TESTED?

After assembly, the ICs are tested and classified as either industrial- or military-grade devices.

After assembly, each individual integrated circuit is put through many tests of its electrical performance to make sure it meets the data-sheet specifications. Just like any electrical circuit, the performance of ICs *changes* with the temperature of the circuit, because the behavior of the individual *components* is affected by temperature. Therefore, each IC type has a temperature-range specification on its data sheet.

In most cases, ICs from one production line are first tested for operation in the temperature range 0 to 70°C; this is satisfactory for most commercial and industrial applications. ICs that pass this testing step are then usually tested in the range —55°C to +125°C; circuits passing tests in this temperature range are suitable for most military and space applications. The two grades thus typically established for each IC type are called *industrial* and *military*.

This system of grading ICs gives industrial and commercial customers a satisfactory product at a lower price than would otherwise be possible. In addition to the wider temperature tolerance demanded, military grades most often have more stringent limits on their electrical performance guarantees as well, of course.

In Chapter 11, we will take a deeper look at integrated circuits, concentrating on digital ICs and their many applications.

Quiz for Chapter 10

1. Why are integrated circuits divided into two categories (digital and linear)?

- ☐ a. These names refer to two general types of packages
- ☐ b. These are the two temperature ranges or grades into which ICs are classified by testing
- ☐ c. ICs are simply circuits that happen to be constructed integrally, and like all circuits, are either switching-type or amplifying-type
- ☐ d. One type processes information and the other controls large working devices such as motors
- ☐ e. ICs are used for either input purposes or output purposes

2. What advantages do ICs have over discrete-device circuits due to their greater complexity (more circuitry in less area)?

- ☐ a. Smaller size
- ☐ b. Higher reliability
- ☐ c. Better overall performance
- ☐ d. Lower cost
- ☐ e. All but c above

3. What are the limitations of integrated circuits?

- ☐ a. Not all electronic components can be efficiently integrated into a silicon chip
- ☐ b. ICs are generally more expensive than their discrete-device counterparts
- ☐ c. ICs are limited to rather low voltages
- ☐ d. ICs can stand a relatively small amount of power dissipation (heating)
- ☐ e. All but b above

4. Which process is used to produce IC semiconductor elements?

- ☐ a. Grown junction
- ☐ b. Alloy junction
- ☐ c. Mesa diffusion
- ☐ d. Planar diffusion
- ☐ e. A unique new process that is limited to ICs

5. Which of the following common electronic components is not found in ordinary ICs?

- ☐ a. Inductors
- ☐ b. Diodes
- ☐ c. Resistors
- ☐ d. Transistors
- ☐ e. Low-capacitance capacitors

6. The components in an ordinary IC chip are electrically isolated from each other by

- ☐ a. A layer of silicon oxide between the components and the substrate
- ☐ b. PN junctions between each component and the P-type substrate. (Conventional current cannot pass from N to P.)
- ☐ c. A layer of undoped silicon, which cannot conduct electricity
- ☐ d. A thin film of photoresist plastic
- ☐ e. They are not isolated, but current runs freely between and among them

7. A resistor in an ordinary IC chip is made by

- ☐ a. Soldering in a discrete resistor
- ☐ b. Depositing a strip of a special resistor material on top of the oxide film
- ☐ c. Using a diode (conducting reverse leakage current) as a resistor
- ☐ d. Diffusing a long, narrow strip of P-region through a larger N-region
- ☐ e. Trimming down a strip of metal interconnection on top of the oxide to a very narrow width

8. What method is used for cutting tiny windows in the oxide layer for diffusing dopant materials into selected areas of the silicon slice?

- ☐ a. A delicate manual technique, using tiny mechanical tools
- ☐ b. The windows occur naturally due to irregular oxidation of the silicon
- ☐ c. Photographically casting tiny shadows on the slice to make holes in a light-sensitive plastic film, followed by an acid bath to etch the oxide under the holes
- ☐ d. A revolutionary secret process not possible with discrete devices
- ☐ e. An electron beam similar to that in a cathode ray tube

9. Why is the plastic dual-in-line package the most popular IC package?

- ☐ a. It is low in cost
- ☐ b. It is easy to insert and solder into printed wiring boards by means of automatic handling equipment
- ☐ c. Its solid construction makes it ruggedly resist vibration and impact
- ☐ d. It is one of the smallest possible packages
- ☐ e. All but d above

10. Why are ICs from one production line classified during the electrical testing step into two grades?

- ☐ a. Circuit performance is always affected by temperature. The military grade works over a wider temperature range than the industrial grade
- ☐ b. Electricity is used for either processing information or doing work
- ☐ c. ICs are either digital or linear types
- ☐ d. Circuit performance is always affected by the frequency of operation. Military types are capable of higher frequencies than the industrial types
- ☐ e. These designations refer simply to different package types

Key Words

Code Converter
Counter
DTL
Data Selector
ECL
Fan-Out
Multiplexer
Negative Logic
Noise Margin
Operating Speed
Positive Logic
Power Dissipation
Read-Only Memory (ROM)
Register
TTL
TTL 54-74 Series

Definitions are found in the glossary in the back of the book.

Digital Integrated Circuits

There are two broad categories of Integrated Circuits. "Linear" ICs contain amplifying-type circuitry, and we'll discuss them in Chapter 13. "Digital" ICs contain switching-type circuitry, and we'll talk about them exclusively in this chapter. If it has been a while since you studied the earlier chapters of this book — on how information is transmitted by digital techniques, how digital decisions are made by means of logic gates, and how digital information is stored by means of flip-flops — better take a few minutes to review Chapters 1, 3, and 4. This chapter is based on those fundamental concepts.

WHAT ARE DIGITAL INTEGRATED CIRCUITS?

Digital ICs are electronic "building blocks" composed of switching circuits that process and store digital information.

Digital ICs are ICs whose basic function is to handle *digital* information by means of switching circuits. You will recall that the other general use for switching circuits is to control power for working devices such as motors. But typical ICs are not yet applicable for this use, because of the limitation to rather low power levels that we discussed in Chapter 10.

HOW ARE DIGITAL INTEGRATED CIRCUITS USED?

Digital ICs are used to *process* information and *store* information in such digital systems as computers, desk calculators, machine tool controls, and frequency-counting instruments. When they use ICs, designers of such systems don't need to spend time and effort in putting transistors together to make gates and flip-flops. They don't even have to put gates and flip-flops together to make more complex circuits. Instead, they can design using larger building blocks, made up of *many* gates and flip-flops, such as the "adder" section of the simple adding machine we discussed in Chapters 3 and 4. Indeed, most of the larger building blocks needed by the designers of digital systems are available in IC form. Consequently, our learning about digital ICs requires becoming acquainted with some of the standard building blocks.

Most digital ICs belong to one of two categories: decision-making circuits or memory circuits.

Digital building blocks can be thought of simply as boxes with inputs and outputs for digital information, plus power-supply and ground connections, as shown in Fig. 11.1. Most digital building blocks available in IC form can be categorized as either *decision-making* or *memory* type. The decision-making type are also called "logic ICs," and they consist mainly of gates; the "memory-type" ICs have flip-flops as their most important components.

Figure 11-1. External Connections of a Digital IC

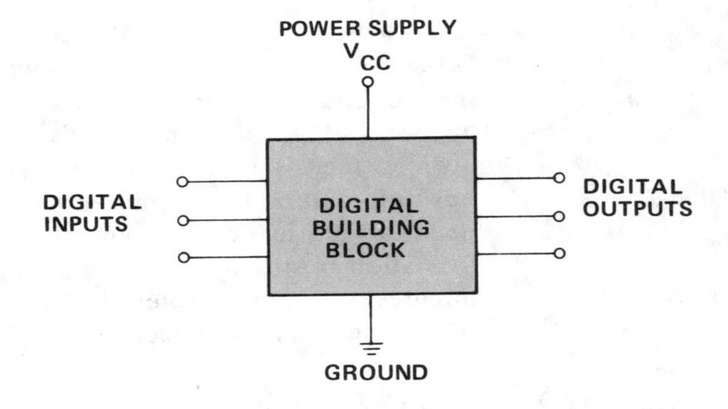

WHAT ARE SOME REPRESENTATIVE DECISION-MAKING BUILDING BLOCKS?

Binary adders, which sum two numbers to obtain a single output, are a common type of decision-making circuit.

One very commonly used building block of the decision-making type is the *adder.* You'll recall that we designed one in Chapter 3. Figure 11.2 outlines the concept of an adder. The inputs are two binary numbers, and the output is the sum of these two numbers in binary form. Clearly, decision-making is involved. The adder contains no flip-flops, as memory building blocks do. It may or may not have a carry output.

Figure 11-2. Binary Adder

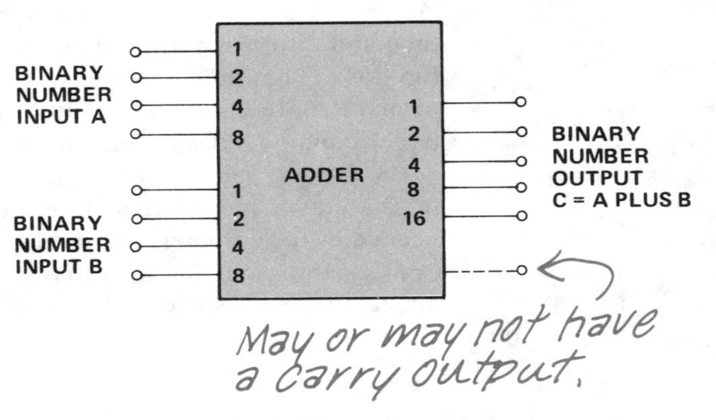

Code converters are decision-making circuits that transform digital information from one format to another (decimal to binary, for example).

Another common decision-making building block is known as a "code converter." You'll recall that the "baby computer" we built in Chapter 4 includes a code converter. Code converters are also called "encoders" and "decoders." In the "sense" section of our baby computer is a block that converts decimal code from the keyboard, to binary code suitable for the adder. Such a code converter is shown in Fig. 11.3.

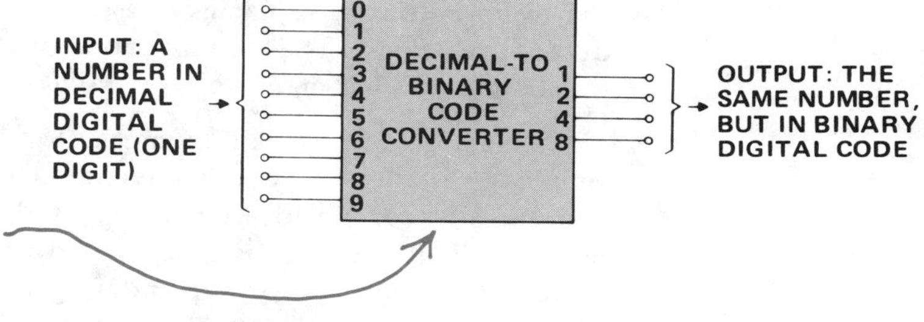

Figure 11-3. Code Converter

Similarly, in the "act" section of our baby computer there appears a block that not only *converts* binary code to proper signals for operating the light display, but also provides power to *drive* the light-emitting diodes.

Data selectors and multiplexers channel information from one device to another within a digital system.

Two more decision-making building blocks are shown in Fig. 11.4. The *data selector* and *multiplexer* are typically used together as a sort of switchboard system for digital information. In this setup, the data selector connects any one of its eight inputs to its one output, upon command from the digital signals received at its "select" inputs. Then, the multiplexer routes information from its one input to any one of its eight outputs, as ordered by inputs to its "select" terminals. Thus, the data selector and multiplexer act like a kind of railroad shifting yard for information signals.

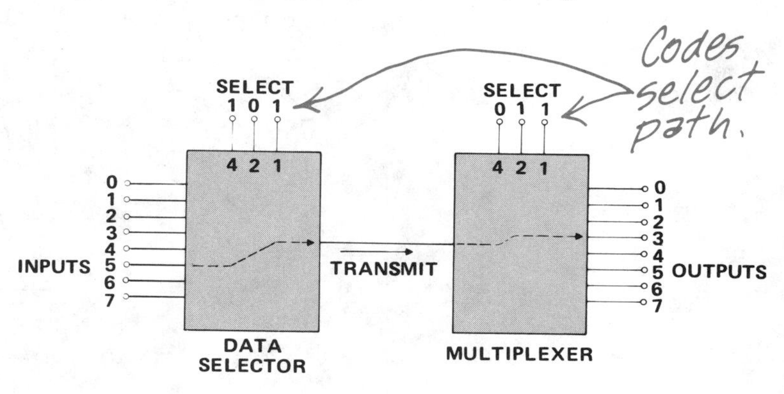

Figure 11-4. Data Selector and Multiplexer

There are, of course, many other kinds of decision-making digital building blocks, but the few we have discussed are among the most popular, and are enough to suggest the almost limitless possibilities.

WHAT ARE SOME REPRESENTATIVE MEMORY - TYPE BUILDING BLOCKS?

Semiconductor memory circuits are composed of flip-flops, which store bits of data by locking their outputs at 1 or 0 as directed by system commands.

Today's digital systems employ various kinds of memory, including magnetic cores, magnetic discs, and magnetic tapes. But the trend is toward replacing these older devices with semiconductor memories. These semiconductor circuits are called "flip-flops," and they are made up essentially of logic gates.

Binary counting makes memories for digital systems relatively simple. All that is needed is a technique for storing logical ones and zeros in certain locations, so that they may be read out as needed. This function is much like writing and reading marks (numbers) on a piece of paper during the manual solution of a long-division problem. The paper does not perform any decision-making — it simply stores the information until you are ready to use it.

A flip-flop is a memory element because it is able to hold, or lock, its output in a particular state when ordered. It flips to a one or flops to a zero when it receives the proper input commands. There are several kinds of flip-flop circuits, but for our example, we will use a simple one called "D-type," shown in Fig. 11.5. The output Q remains constant so long as the clock input is logical zero. But when the clock is pulsed — changed to one and back to zero — this causes the output to repeat whatever state is at the data input D at that moment, and hold it until the next clock pulse.

Where there is need in a system to store *groups* of bits ("words"), several flip-flops are used under some sort of unified

Figure 11-5. D Flip-Flop

Registers consist of several flip-flops controlled as a unit to store binary words (groups of bits).

control. Such a combination is called a "register." In Chapter 4, we discussed a variety of register called a "shift register," in which the output of each flip-flop feeds the input of another in a series. In the shift register, when all the flip-flops are clocked simultaneously, bits of data shift step by step along the register, from the input of the first flip-flop to the output of the last.

Another type of register, in which several bits can be put in or taken out *simultaneously*, is called a *parallel* register. The memory block in our "baby computer" of Chapter 4 is a parallel combination of four two-bit shift registers.

A counter is a building block that keeps a running total of the digital pulses received at its single input.

Still another common memory building block is the *counter*. A typical counter is shown in Fig. 11.6. This block simply counts digital pulses received at the single input and *remembers* the cumulative total. The total is then presented at the four outputs as a four-bit binary number. A portion of the count sequence is shown in the figure. This particular counter is designed so that when the total reaches 1111 (15), it reverts to 0000 on the next pulse. Counters can be designed to revert to zero after reaching virtually any number. A counter that reverts to zero after a total of *nine* is called a "decade" counter. A counter that goes back to zero after reaching *eleven* would be called a "divide-by-twelve" counter, etc. Some counters will count backward as well as forward, and these are called "up-down" counters.

Read-only memories (ROMs) are digital building blocks that contain permanently stored, unchangeable information.

Another type of building block that is usually classified as memory-type is the *read-only memory* (ROM). In a ROM, information is stored in a permanent form during the manufacture of the circuit, chiefly by the design of the interconnections. "Read-only memory" means that the information can *only* be read out, and that it cannot be written in, or changed. One popular use for the ROM is as a character generator.

Figure 11-6. Digital Counter

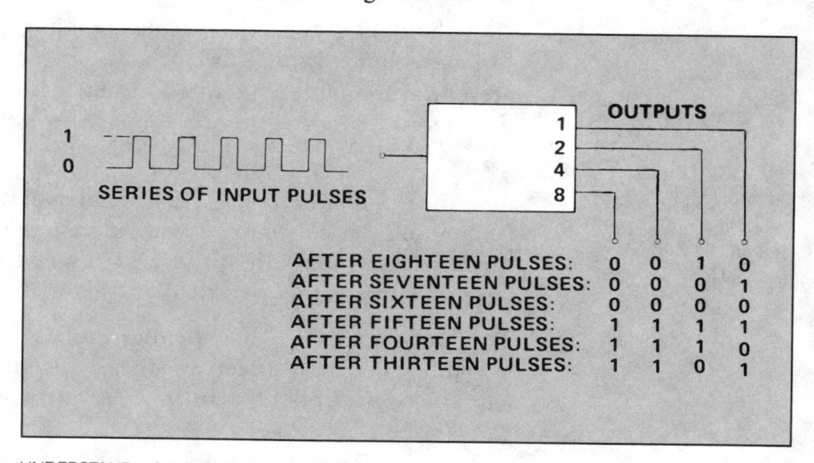

The storing of data for character generation is a common ROM application.

The ROM shown in Fig. 11.7 is a particular memory that contains 2,240 bits of information arranged as 64 words (groups of bits) of 35 bits each. Any one of the words can be called up at the thirty-five output lines by applying the proper binary number (from zero to 63) at the six inputs. Each of the 35 outputs controls one lamp in the five-by-seven array of 35 lamps.

Figure 11-7. Character Generator

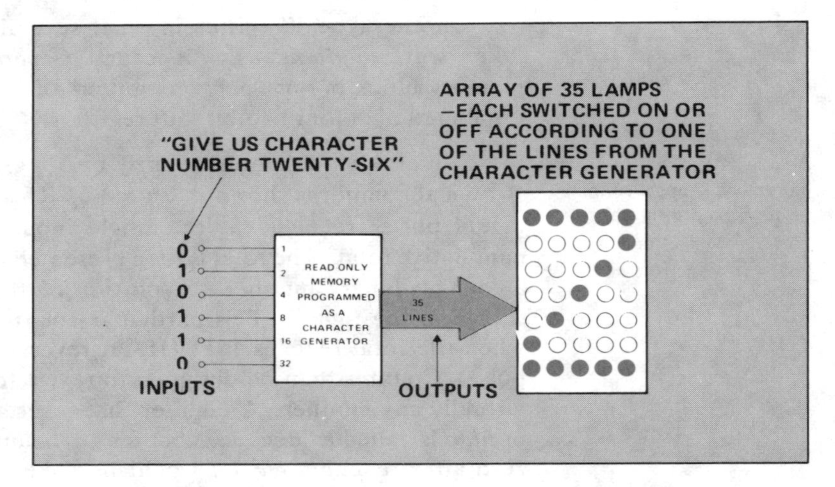

In Fig. 11.7, the input command has called up the word stored in location number 26, and this word produces the image of the character "Z" in the lamp array. In this manner, each word produces a different character (a letter, a number, a punctuation mark, etc.) on command — hence the name, "character generator."

WHAT SPECIFICATIONS ARE FOUND ON DIGITAL IC DATA SHEETS?

The data sheets of all digital building blocks available in IC form are quite similar in format. Figures 11.8 through 11.15 present a complete integrated circuit data sheet for an eight-bit shift register.

Digital IC data sheets contain information of two kinds: functional specifications and electrical specifications.

The important specifications are divided into two classes: *functional* and *electrical.* Functional specifications tell what the IC building block *does* — what decisions it makes. Electrical specifications, much like those of discrete devices, describe the circuit in terms of its electrical characteristics.

Since *functional* specifications tell us what the IC does with the information it receives, these specifications are normally expressed in terms of a *truth table,* with which you're familiar.

Functional specifications commonly take the form of truth tables.

While you are familiar with truth tables that use one and zero, you should be aware that there is a recent trend to use the letters "H" and "L" in place of one and zero. In discussing truth tables, you may recall that we said high voltage *usually* stands for one and low voltage usually stands for zero. But it's up to the designer, after all, to decide which means which. Most designers let the most positive voltage stand for one and the least positive voltage stand for a zero; this is called "positive" logic, and is by far the more popular. But there are some systems in which the least negative voltage stands for zero and the most negative voltage stands for a one — an alternative known as "negative" logic. So, a truth table using "H" and "L" satisfies both.

But note this warning. Using positive logic for a typical building block produces an entirely *different* truth table than if negative logic were chosen. And that means the building block performs a completely *different function.* You can prove this to yourself by taking the truth table for the AND gate that we showed as an example, and changing all the ones to zeros and all the zeros to ones. The truth table that results is the truth table for an OR function. To summarize the changes, a gate that performs the AND function using positive logic will act as an OR gate, using negative logic. Similarly, a positive OR gate becomes a negative AND gate. So a complete functional specification includes the designation "positive" or "negative." As you've probably recognized, an inverter is always an inverter.

Electrical specifications common to all digital ICs include noise margin, fan-out, operating speed, and power dissipation.

Electrical specifications of a digital building block include some with which we are already familiar. But in addition, there are five major specifications that we might call "universal digital IC specifications" because they apply to all digital building blocks:

Noise margin describes how securely a building block transmits and receives correct information in spite of "noise." The ability of a transmitter-receiver combination to discriminate between noise and real data depends mainly on the "safety margin" between the voltage (for the two logic states) *produced* by the transmitting output, and that *required* by the receiving input.

These voltages are specified on the data sheet as V_{OH}, V_{OL}, V_{IH}, and V_{IL}. In these symbols, "O" = output, "I" = input, "H" = high, and "L" = low. Noise margin, although it is not usually specified separately on the data sheet, is simply V_{OH} minus V_{IH}, and V_{IL} minus V_{OL}. The output voltages here apply to the *transmitting* block, and the input voltages apply to the *receiving* block. For example, if both blocks are circuits from the 54/74 TTL

Figure 11-8.
IC Data Sheet

TTL
MSI

TYPES SN54165, SN54LS165, SN74165, SN74LS165
PARALLEL-LOAD 8-BIT SHIFT REGISTERS

BULLETIN NO. DL-S 7611375, OCTOBER 1976

- Complementary Outputs
- Direct Overriding Load (Data) Inputs
- Gated Clock Inputs
- Parallel-to-Serial Data Conversion

TYPE	TYPICAL MAXIMUM CLOCK FREQUENCY	TYPICAL POWER DISSIPATION
'165	26 MHz	210 mW
'LS165	35 MHz	105 mW

SN54165, SN54LS165 . . . J OR W PACKAGE
SN74165, SN74LS165 . . . J OR N PACKAGE
(TOP VIEW)

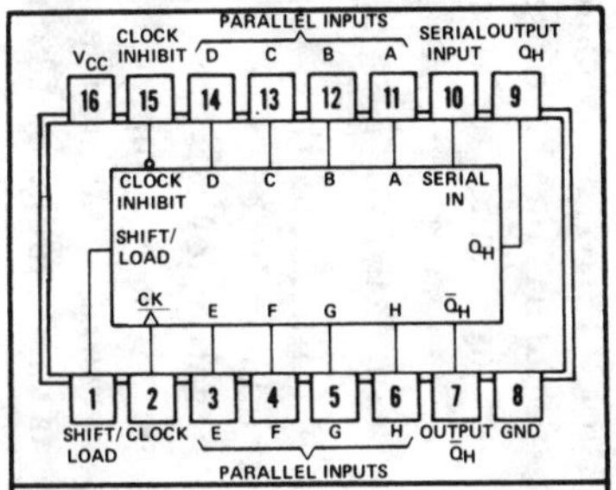

V_{CC} 16, CLOCK INHIBIT 15, PARALLEL INPUTS D 14, C 13, B 12, A 11, SERIAL INPUT 10, OUTPUT Q_H 9
SHIFT/LOAD 1, CLOCK 2, PARALLEL INPUTS E 3, F 4, G 5, H 6, OUTPUT $\bar{Q}_H$ 7, GND 8

CLOCK INHIBIT, SHIFT/LOAD, CK, D, C, B, A, SERIAL IN, E, F, G, H, Q_H, $\bar{Q}_H$

positive logic: see description

description

The '165 and 'LS165 are 8-bit serial shift registers that shift the data in the direction of Q_A toward Q_H when clocked. Parallel-in access to each stage is made available by eight individual direct data inputs that are enabled by a low level at the shift/load input. These registers also feature gated clock inputs and complementary outputs from the eighth bit. All inputs are diode-clamped to minimize transmission-line effects, thereby simplifying system design.

**Figure 11-9.
IC Data Sheet
(Continued)**

TYPES SN54165, SN74165
PARALLEL-LOAD 8-BIT SHIFT REGISTERS

Clocking is accomplished through a 2-input positive-NOR gate, permitting one input to be used as a clock-inhibit function. Holding either of the clock inputs high inhibits clocking and holding either clock input low with the shift/load input high enables the other clock input. The clock-inhibit input should be changed to the high level only while the clock input is high. Parallel loading is inhibited as long as the shift/load input is high. Data at the parallel inputs are loaded directly into the register on a high-to-low transition of the shift/load input independently of the levels of the clock, clock inhibit, or serial inputs.

recommended operating conditions

	SN54165			SN74165			UNIT
	MIN	NOM	MAX	MIN	NOM	MAX	
Supply voltage, V_{CC}	4.5	5	5.5	4.75	5	5.25	V
High-level output current, I_{OH}			−800			−800	μA
Low-level output current, I_{OL}			16			16	mA
Clock frequency, f_{clock}	0		20	0		20	MHz
Width of clock input pulse, $t_{w(clock)}$	25			25			ns
Width of load input pulse, $t_{w(load)}$	15			15			ns
Clock-enable setup time, t_{su} (see Figure 1)	30			30			ns
Parallel input setup time, t_{su} (see Figure 1)	10			10			ns
Serial input setup time, t_{su} (see Figure 2)	20			20			ns
Shift setup time, t_{su} (see Figure 2)	45			45			ns
Hold time at any input, t_h	0			0			ns
Operating free-air temperature, T_A	−55		125	0		70	°C

Figure 11-10. IC Data Sheet (Continued)

electrical characteristics over recommended operating free-air temperature range (unless otherwise noted)

PARAMETER			TEST CONDITIONS†	SN54165			SN74165			UNIT
				MIN	TYP‡	MAX	MIN	TYP‡	MAX	
V_{IH}	High-level input voltage			2			2			V
V_{IL}	Low-level input voltage					0.8			0.8	V
V_{IK}	Input clamp voltage		V_{CC} = MIN, I_I = −12 mA			−1.5			−1.5	V
V_{OH}	High-level output voltage		V_{CC} = MIN, V_{IH} = 2 V, V_{IL} = 0.8 V, I_{OH} = −800 μA	2.4	3.4		2.4	3.4		V
V_{OL}	Low-level output voltage		V_{CC} = MIN, V_{IH} = 2 V, V_{IL} = 0.8 V, I_{OL} = 16 mA		0.2	0.4		0.2	0.4	V
I_I	Input current at maximum input voltage		V_{CC} = MAX, V_I = 5.5 V			1			1	mA
I_{IH}	High-level input current	Shift/load	V_{CC} = MAX, V_I = 2.4 V			80			80	μA
		Other inputs				40			40	
I_{IL}	Low-level input current	Shift/load	V_{CC} = MAX, V_I = 0.4 V			−3.2			−3.2	mA
		Other inputs				−1.6			−1.6	
I_{OS}	Short-circuit output current§		V_{CC} = MAX	−20		−55	−18		−55	mA
I_{CC}	Supply current		V_{CC} = MAX, See Note 3		42	63		42	63	mA

NOTE 3: With the outputs open, clock inhibit and clock at 4.5 V, and a clock pulse applied to the shift/load input, I_{CC} is measured first with the parallel inputs at 4.5 V, then with the parallel inputs grounded.

†For conditions shown as MIN or MAX, use the appropriate value specified under recommended operating conditions.

‡All typical values are at V_{CC} = 5 V, T_A = 25°C.

§Not more than one output should be shorted at a time.

Figure 11-11. IC Data Sheet (Continued)

switching characteristics, $V_{CC} = 5\ V$, $T_A = 25°C$

PARAMETER¶	FROM (INPUT)	TO (OUTPUT)	TEST CONDITIONS	MIN	TYP	MAX	UNIT
f_{max}			$C_L = 15\ pF$, $R_L = 400\ \Omega$, See figures 1 thru 3	20	26		MHz
t_{PLH}	Load	Any			21	31	ns
t_{PHL}					27	40	
t_{PLH}	Clock	Any			16	24	ns
t_{PHL}					21	31	
t_{PLH}	H	Q_H			11	17	ns
t_{PHL}					24	36	
t_{PLH}	H	$\bar{Q}_H$			18	27	ns
t_{PHL}					18	27	

¶f_{max} ≡ maximum clock frequency
t_{PLH} ≡ propagation delay time, low-to-high-level output
t_{PHL} ≡ propagation delay time, high-to-low-level output

Figure 11-12. IC Data Sheet (Continued)

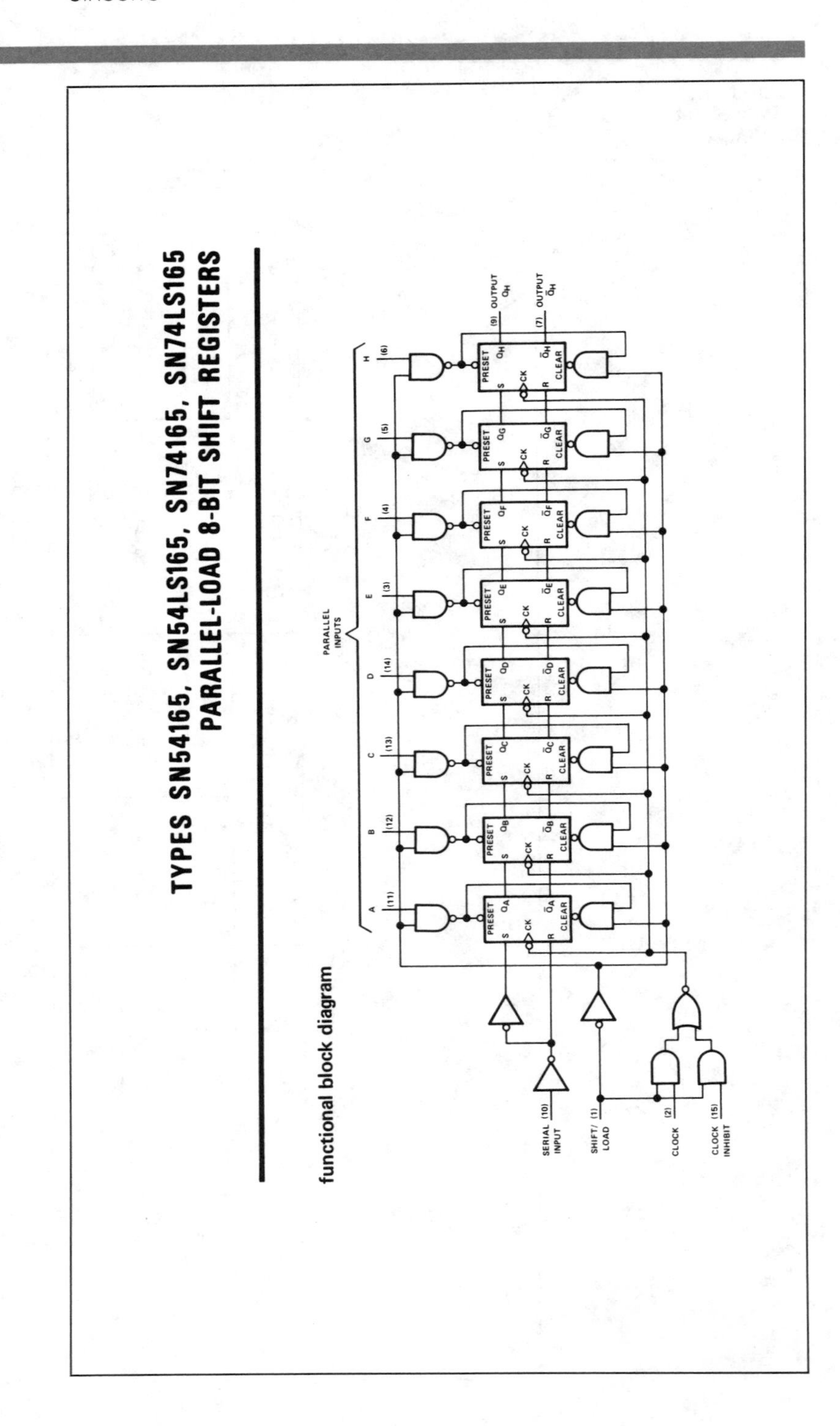

**Figure 11-13.
IC Data Sheet
(Continued)**

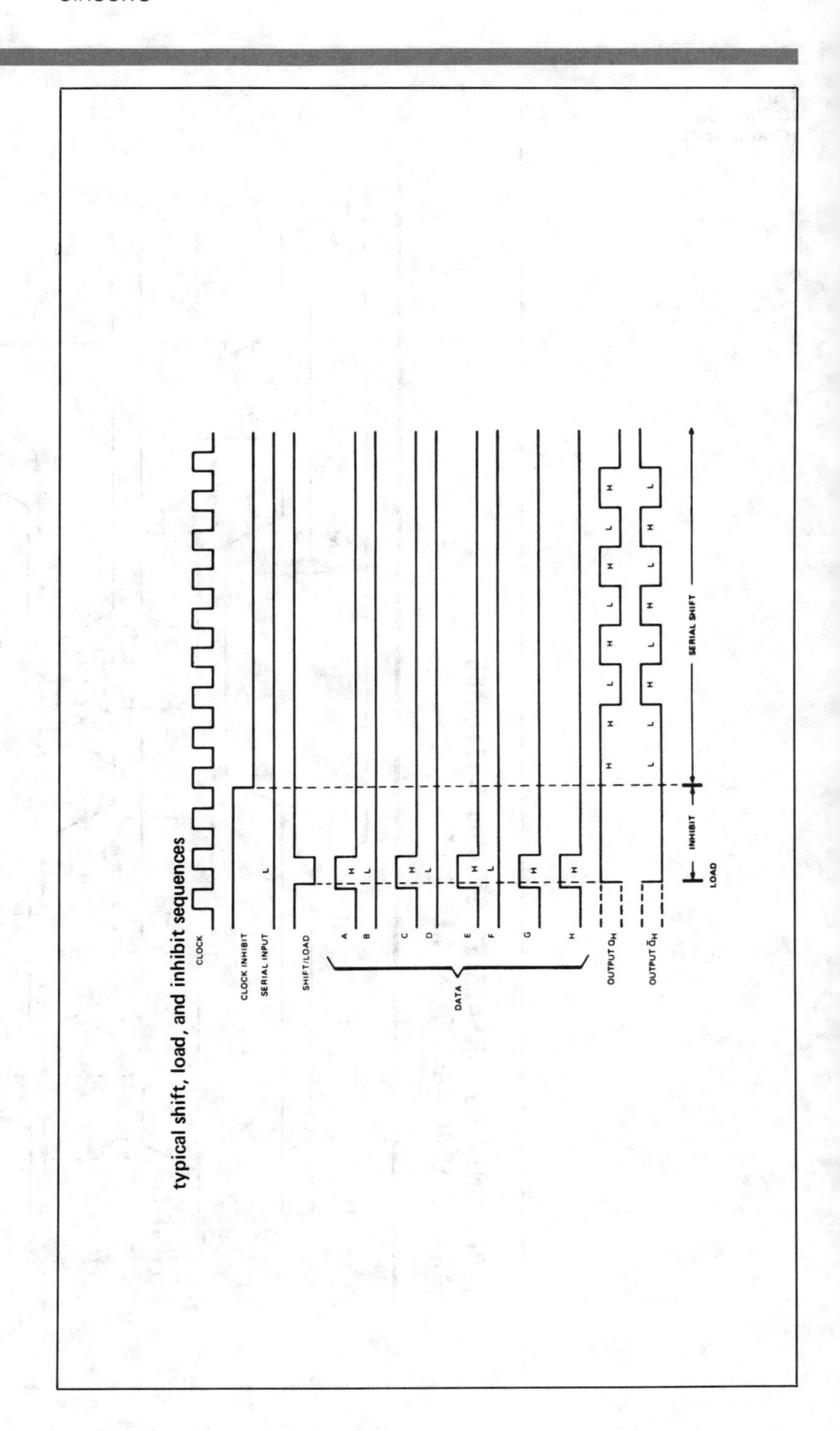

Figure 11-14. IC Data Sheet (Continued)

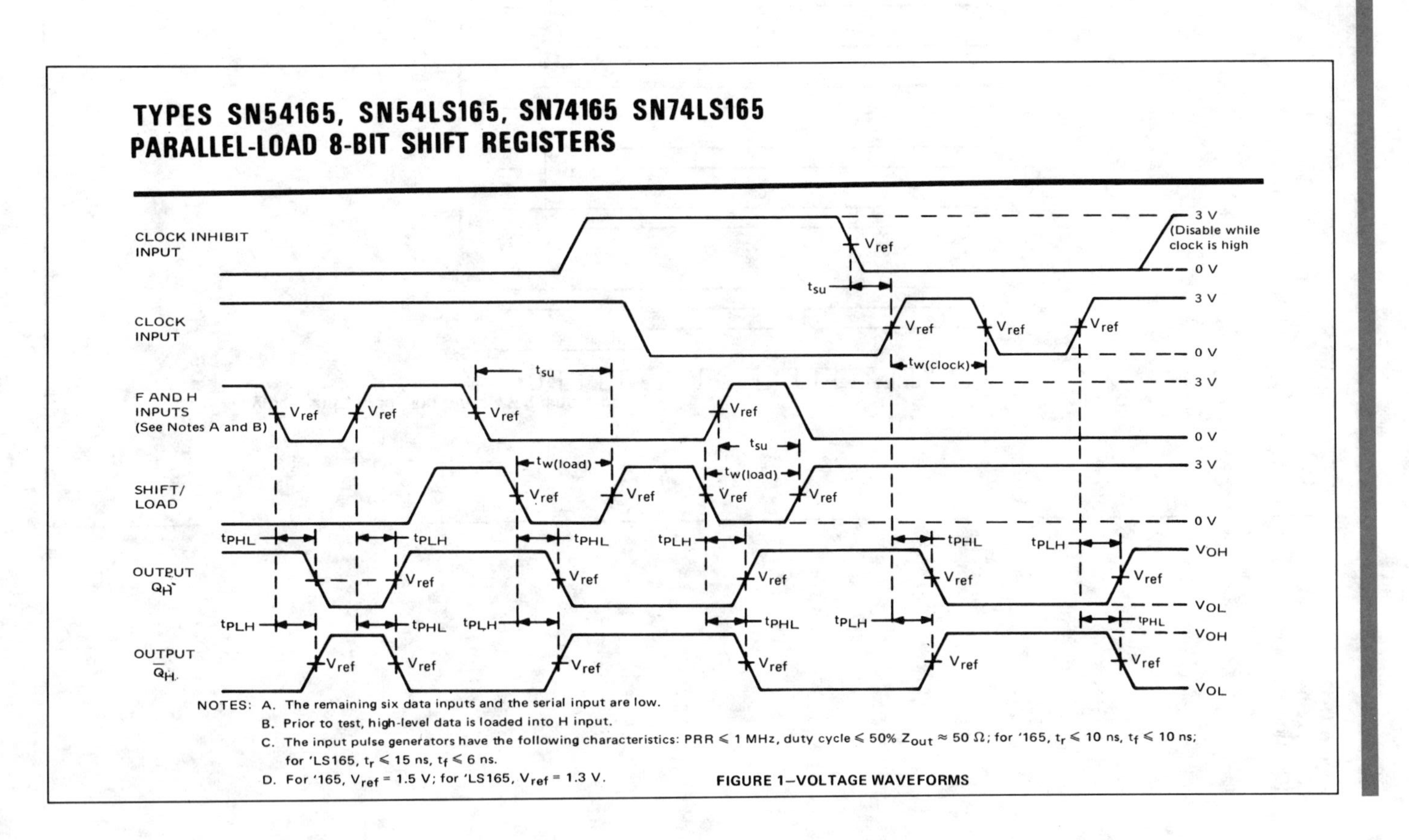

**Figure 11-15.
IC Data Sheet
(Continued)**

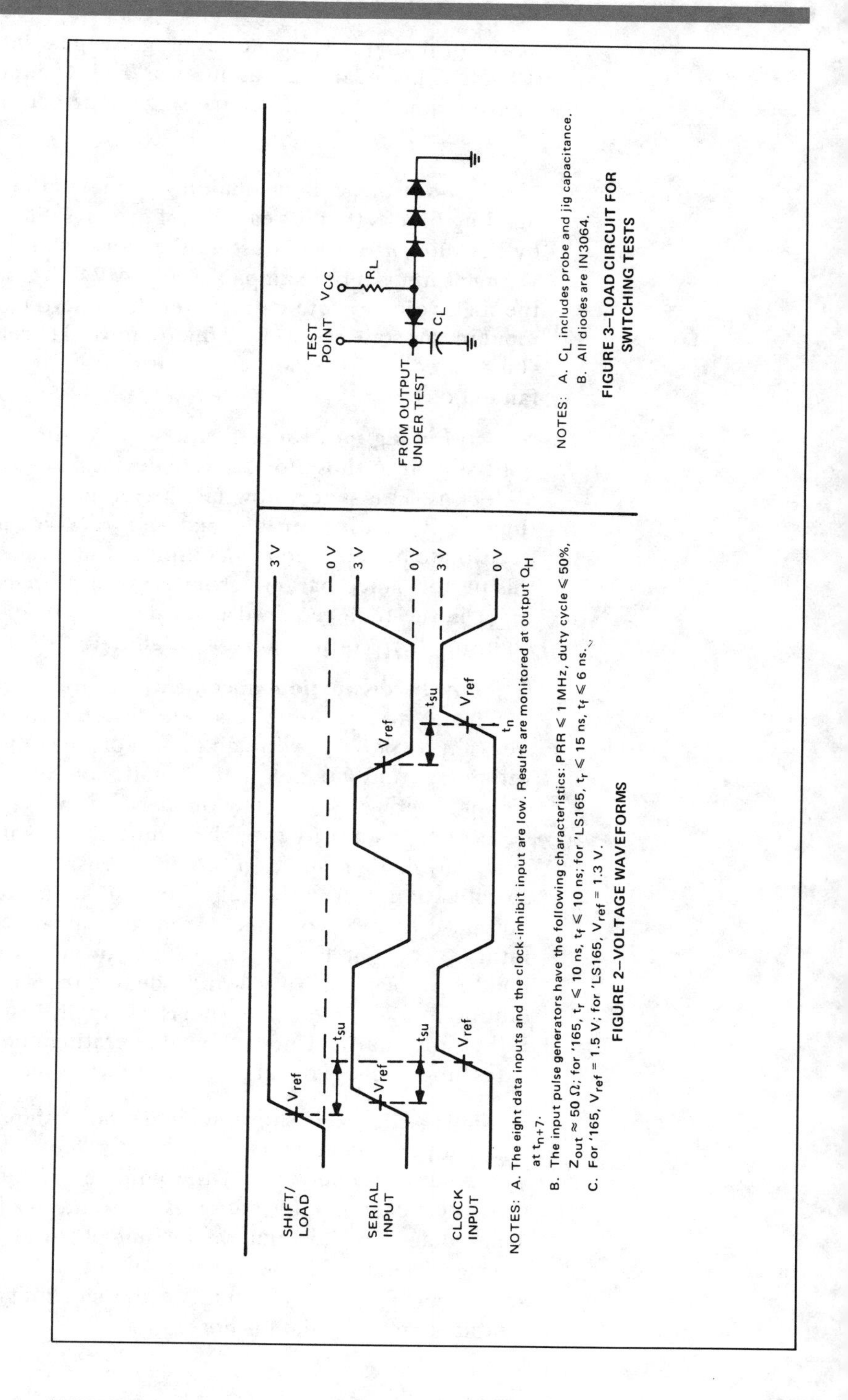

family of digital ICs, the noise margin is guaranteed to be at least 0.4 volt. This means that as much as 0.4 volt of noise can mess up a transmission link, without causing incorrect information to be received.

Fan-out of a digital building block is the number of other building blocks that it's capable of driving. Fan-out is determined by the *current* ratings, just as the noise margin depends on the *voltage* ratings. For example, in the 54/74 TTL series, an output in the high-voltage state can deliver 400 microamps of current. An input in this state requires 40 microamps. Therefore, the fan-out is 400 divided by 40 — so the output can drive 10 inputs, and has a fan-out of 10.

Operating speed specifications for digital building blocks are quite similar to those for discrete devices. The operating speed of a block is, in essence, how fast it can make a decision. It is the time the IC takes, after information is received at an input, for the decision to be reported at an output. IC data sheets usually specify this performance parameter in terms of *propagation delay times:* t_{PLH} is the time required for a decision to change from low to high, and t_{PHL} is the time for a high-to-low decision.

Power dissipation specifications for digital blocks differ somewhat from those for discrete devices. In the case of devices, we want to know how much power or heat the unit can *tolerate* without getting so hot that it malfunctions. Now a digital IC dissipates power just as any other electronic component does, but we normally want to know how much power it *uses up,* not how much it can *stand.* In other words, the power dissipation specification for an IC tells how much power it consumes in normal operation, by conversion to heat, much like the wattage rating of a light bulb. Since the power dissipation of a given building block can vary widely depending on its operating conditions at the time, data sheets normally list this specification rather informally. Under typical operating conditions, 54/74 TTL gates might dissipate about 10 milliwatts per gate.

Timing Diagrams show the time relationships between the input and output signals for the circuit. These further define the operating characteristics of the unit. In the case of the SN74165 shift register, this information is presented in Fig. 11.2 and 11.14. These diagrams show that the loading of parallel data into the shift register occurs when the shift-load input makes a high-to-low transition, regardless of the state of the clock signal. When a clock inhibit is high, the data is not shifted serially. When clock inhibit is low, the data in the register is shifted out serially each time the clock makes a high-to-low transition. Much of this information is contained in the description portion of Fig. 11.8. The timing waveforms show the interrelationships between signals and show at a glance the way the circuit operates.

HOW DO DIGITAL INTEGRATED CIRCUITS WORK?

Digital ICs contain gates and flip-flops, arranged as needed to implement the logical functions desired.

Digital ICs work by combining gates and flip-flops (which are essentially made up of gates) in a logical arrangement to produce a desired output, given an appropriate input. So if you know how various gate circuits operate, you have a good insight into how ICs operate. We have already covered the three basic gates: AND, OR, and Invert. We have also covered the two most common modifications of the basic circuits; the NAND and NOR gates. The art of integrated circuitry consists in connecting a selection of these gates together as necessary to perform the required logical function, and then reproducing the electrical components and interconnections of these gates in a single chip of silicon.

A number of different circuit variations can be used to obtain a given logical function.

Any number of different circuit arrangements can function as logic gates. In Chapter 5, for example, you saw how diodes can perform AND and OR functions, and how diode gates can be followed by a transistor to add power to the output. In Fig. 11.16, you see a simple DTL (diode-transistor logic) gate that we built in Chapter 5. This is a *positive* logic NAND gate.

This DTL circuit provides a general idea of how gate circuits work. A "one" or most positive voltage (H) at both inputs supplies base current to turn the transistor on. This then connects the output directly to ground, for a "zero" or least positive (L) output.

Figure 11-16.
DTL NAND Gate

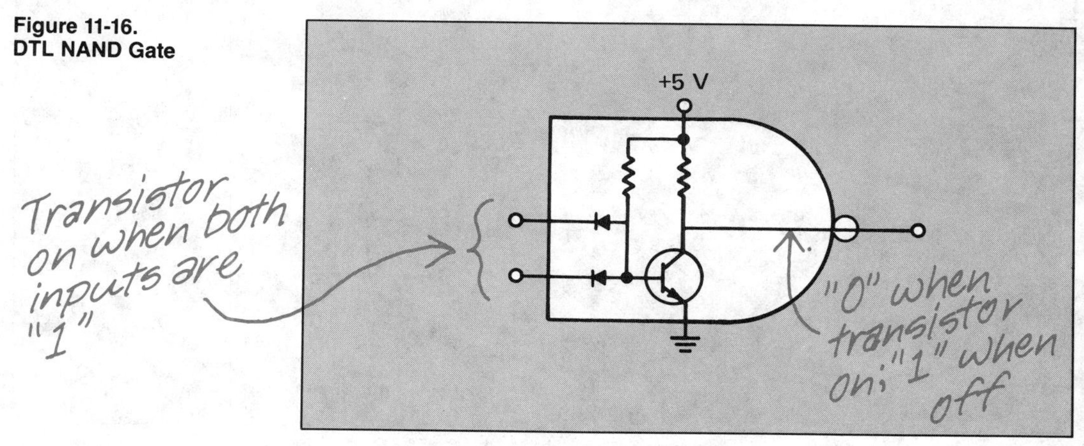

But when neither input is high, the base is clamped to ground by the least positive ''zero'' voltage, the transistor stays off, and the resistor connected to the power supply provides a ''one'' output — a most positive voltage ''high'' output.

Notice the outlined figure around this circuit. The semi-circle shows that the circuit is an AND gate. And the circle at its apex signifies negation. So this circuit is identified as a NAND gate.

Of the various logic circuit technologies that have been developed, TTL (transistor-transistor logic) has become the most popular.

DTL logic and the older types of logic you may have heard of — RTL, RCTL, DCTL, DTL, and CTL — have pretty well passed into history. System designers have largely abandoned them in favor of today's most popular logic type, TTL. "TTL" simply stands for "transistor-transistor logic."

Figure 11.17 shows a positive NAND gate that uses TTL logic. Notice that the shaded area indicates it is an AND gate and the output is inverted — making it a NAND gate.

The input transistor is an NPN type, with a *separate* emitter region for each input. Thus, it functions essentially as three diodes, as indicated by the small figure. Working together with the resistor, it acts as a positive AND gate. That is, power-supply current flows out of its output only as long as both inputs are "high." The output of this internal gate is inverted and amplified

Figure 11-17. TTL NAND Gate

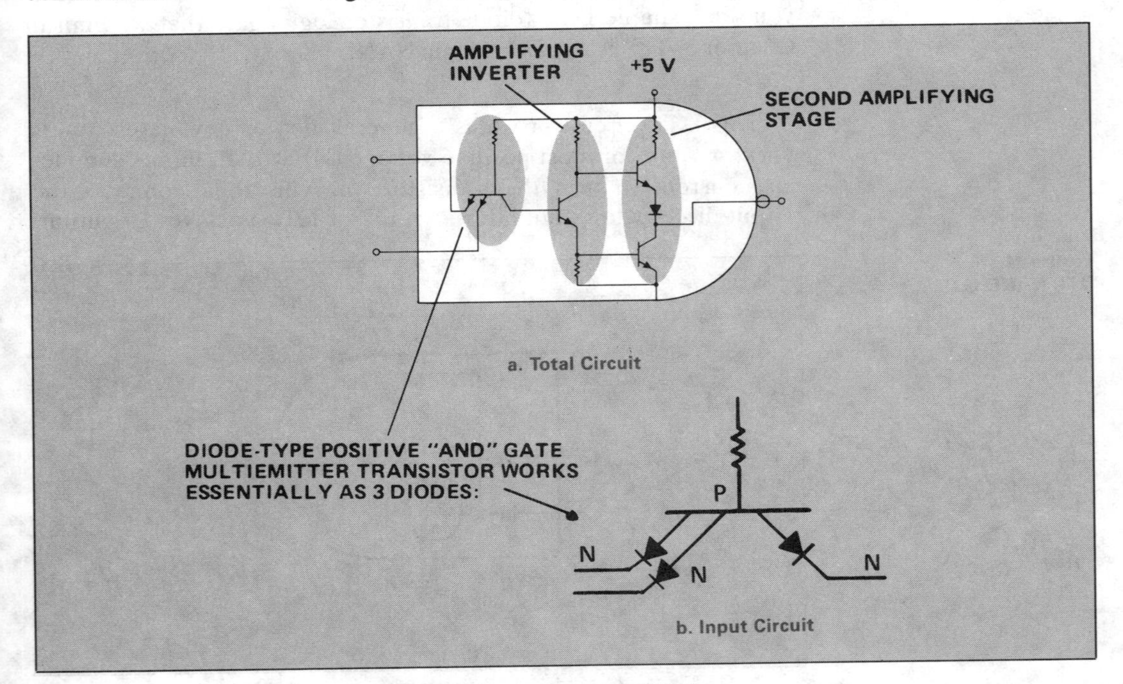

by the remainder of the circuitry, resulting in a positive NAND function at the output.

TTL logic of this kind has become greatly popular largely because of its excellent noise margin, high fan-out, fast operating speed, and low power dissipation.

Most TTL digital building blocks are identified as members of the "54/74 Series." The "54" means the device is suitable for operation in the *military* temperature range; "74" means it is suitable for the *industrial* temperature range, which is not as severe. Immediately following these two digits, a letter may show that the IC is a modification of the standard device type (which has no letter): an "L" means the device is designed for *low* power dissipation; an "H" means it is designed for higher operating speed; and an "S" means that special diodes called "Schottky barrier diodes" have been incorporated into the circuitry to increase the operating speed — at the cost of little, if any increase in power dissipation.

Our sample data sheet includes type number "SN54165." The "SN" is simply the prefix for all integrated circuits. The "54," as we have said, indicates the device is suitable for the military temperature range of —55°C to 125°C. The fact that there is no letter following the "54" indicates that it is a *standard* circuit. The "165" is simply the sequential number of this circuit in the series.

Although TTL offers the most optimal combination of desirable circuit characteristics, ECL (emitter-coupled logic) is sometimes used in very large computers, where its greater speed outweighs its drawbacks.

Its *combination* of excellent characteristics is what has made TTL logic the overwhelming favorite of system designers at this writing. And the 54/74 Series boasts the second-highest operating speed of all logic types. But there is a logic type that is even faster — ECL, or emitter-coupled logic. Yet ECL pays for its fast operating speed in terms of a large increase in power dissipation, and its noise margin and fan-out are not considered as good as those of TTL. At this time, ECL is used almost exclusively in very large computers where the higher speed makes the other sacrifices worthwhile.

Our survey of digital integrated circuits is now as complete as the size of this book permits. In the next chapter, we will talk about MOSFET (metal-oxide-semiconductor field-effect transistor) integrated circuits, then finish up with a discussion of linear integrated circuits.

Quiz for Chapter 11

1. Why are virtually all switching-type integrated circuits called "digital"?

- ☐ a. They are about the size and shape of a digit or finger
- ☐ b. They process digital information
- ☐ c. They switch large working devices on and off
- ☐ d. "Di" in "digital" means "two," referring to the two purposes of electricity in systems — processing information and doing work
- ☐ e. All but b above

2. Digital electronic systems, from our simple adding machine of Chapter 4 to large computers, operate by means of:

- ☐ a. Amplifying-type circuits making analog decisions
- ☐ b. Logic gates making decisions using digital information
- ☐ c. Flip-flops (and various types of non-semiconductor memory units) storing digital information
- ☐ d. Mechanical levers and cams operated in turn by electricity
- ☐ e. b and c above

3. Digital electronic systems are designed by putting together building blocks available in integrated-circuit form which can be classified as:

- ☐ a. Processing information or doing work
- ☐ b. Positive or negative logic
- ☐ c. Decision-making (or "logic") type, consisting mainly of gates; or memory-type, containing principally flip-flops
- ☐ d. Functional and electrical
- ☐ e. Switching and amplifying

4. The meaning of the letter symbols "H" and "L" in functional descriptions (truth tables) of digital building blocks is:

☐ a. Higher voltage and lower voltage, referring to the two binary digital signal levels
☐ b. Higher power and lower power, referring to these two classifications of semiconductor devices
☐ c. High frequency and low frequency
☐ d. High current and low current
☐ e. High speed and low speed

5. If a certain circuit acts as an OR gate when used with negative logic (H = 0, L = 1), what function will it perform when used with positive logic (H = 1, L = 0)?

☐ a. AND
☐ b. OR
☐ c. NOT
☐ d. NAND
☐ e. NOR

6. Noise margin, which (by being small) is one indication of how likely it is that information communicated between digital building blocks will be incorrect due to noise, depends on:

☐ a. Output current capabilities and input current requirements
☐ b. The "safety margin" between the output voltage produced by the transmitting block and input voltage required by the receiving block for each of the two logic states
☐ c. Output power and required input power for the two logic states
☐ d. The "safety margin" between the noise level and the noise figure
☐ e. The noise figure (NF)

7. The fan-out capability of a digital building block depends on the current capability of its output and the current requirement of each input driven by that output, and may be defined as

- ☐ a. The number of other inputs that can transmit to one input
- ☐ b. The maximum power dissipation (heat generation) that the unit can stand
- ☐ c. The amount of cooling (fanning the heat out) required
- ☐ d. The number of inputs that one output can transmit to
- ☐ e. b and c above

8. "Operating speed" is the general name for the speed capability of any semiconductor device or circuit. In the data sheet of a digital building block, it is typically expressed in terms of

- ☐ a. Transition frequency f_T.
- ☐ b. Capacitance C
- ☐ c. Miles per hour, or centimeters per second
- ☐ d. Propagation delay times for both possible output transitions: t_{PHL} and t_{PLH}
- ☐ e. The "warm-up" time delay after the power is turned on until normal operation is possible

9. The power dissipation specification of most *discrete* semiconductor devices is the absolute maximum heat production the device is guaranteed to withstand. But the power dissipation specification normally discussed for digital building blocks (usually not formally guaranteed on the data sheet) is

- ☐ a. The same as that for most discrete semiconductor devices
- ☐ b. The amount of power consumed and converted to heat in normal operation
- ☐ c. A minimum limit required to maintain high enough operating temperatures
- ☐ d. Fan-out multiplied by noise margin
- ☐ e. Really not a matter of power dissipation at all

10. TTL (transistor-transistor logic), DTL (diode-transistor logic), and ECL (emitter-coupled logic), which are frequently used to refer to certain "families" of digital integrated circuits, are actually names of

- ☐ a. Companies that originated the families
- ☐ b. Alternatives to positive and negative logic
- ☐ c. Varieties of positive and negative logic
- ☐ d. Special semiconductor devices similar to transistors
- ☐ e. General varieties of electronic circuits used as logic gates, from which, in essence, the building blocks in each series are constructed

12 MOS AND LARGE-SCALE INTEGRATED CIRCUITS

Key Words

Bipolar
Drain
FET
Gate
Junction-FET
Microcomputer
Microprocessor
MOS
RAM
SSI, MIS, LIS, VLSI
Source

Definitions are found in the glossary in the back of the book.

MOS and Large-Scale Integrated Circuits

The Integrated Circuits we have seen so far are all "bipolar," because they all use bipolar transistors — the only kind we've studied. MOS integrated circuits, however, use field-effect transistors (FETs), a different kind of structure.

WHAT ARE MOS INTEGRATED CIRCUITS?

MOS (metal-oxide-semiconductor) ICs are made almost entirely from FETs (field-effect transistors), which allow much denser circuitry than do bipolar circuit elements.

MOS ICs are digital ICs whose circuitry used MOSFETs (metal-oxide-semiconductor field-effect transistors) — and virtually no resistors, diodes, bipolar transistors, or other components. MOS ICs are already widely used at this writing, and their popularity is increasing rapidly, because MOSFET elements permit greater *complexity* than bipolar elements. This means more *circuitry,* and hence more gates and flip-flops, in the same *area* of a semiconductor chip. This ability to jam more functions into a tiny piece of silicon "real estate" is the key to greater economy in integrated circuits. Indeed, this continuing trend toward greater and greater complexity is the key to progress in electronics.

To get a rough idea of the complexity made possible by modern integrated circuit techniques, take a look at Fig. 12.1. This is a rather large MOS integrated circuit — it's almost a quarter of an inch square. It would only take twelve of these to cover a postage stamp. Still, it's quite remarkable because it contains the equivalent of about 6,000 discrete devices on a single chip. Don't worry about the areas that are identified in this photo — suffice it to say that this is a *computer on a chip.*

You can take this MOS chip, hook on not too much additional hardware, and build a complete computer in a package about the size of a breadbox. And this little breadbox-size computer would have all the power — all the ability to solve complex problems — that a room-size computer had just a few

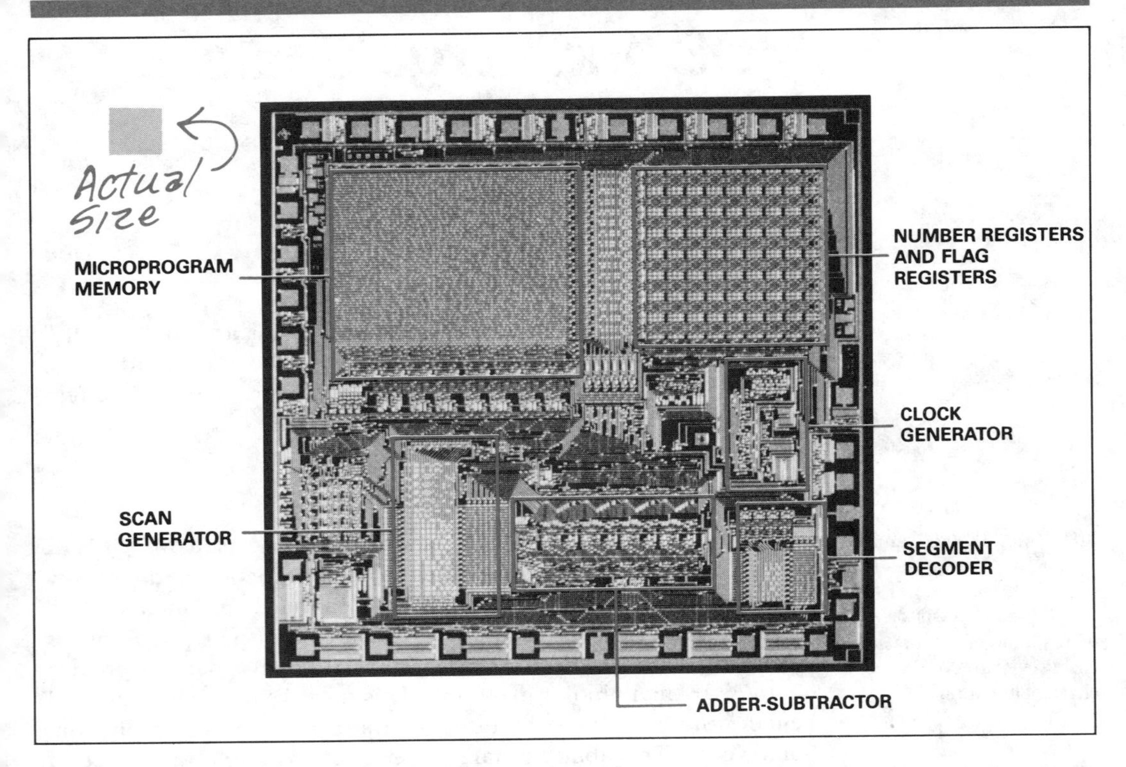

Figure 12-1. MOS Integrated Circuit Chip

years ago! *Bipolar* integrated circuits are fantastic enough — but MOS technology packs several times as much complexity into the same size chip.

HOW DO MOSFETS WORK?

This typical MOSFET design has an N-type substrate with two P regions ("source" and "drain"), which are covered by a layer of silicon oxide and connected to the outside by metal strips. A third metal strip (the "gate") lies atop the oxide, above the gap between source and drain.

A conceptual cross-section of the type of MOSFET used in MOS ICs is shown in Fig. 12.2. Two P-regions, called the *source* and the *drain,* are diffused side by side into the surface of an N-type silicon slice. A layer of insulating silicon oxide is grown over the surface. Two strips of metallization are made to penetrate through windows in the oxide, to contact the silicon. A third metal strip, the *gate,* lies on top of the oxide, over the gap between the P regions.

The source, gate, and drain correspond to the emitter, base, and collector of a bipolar transistor.

These parts of a MOSFET perform much the same *function* in a circuit as the corresponding parts of a PNP transistor. The *source acts as emitter,* the *gate* acts as *base*, and the *drain* acts as *collector.* When no control signal is applied to the gate, working current—which is trying to pass from source to drain — cannot flow. The source, substrate, and drain in essence work like a PNP transistor — they block working current unless control current (conventional current) is withdrawn from the N region. But in the MOSFET, current is instead trying to flow *into* the N region from the source metallization through a window provided in the oxide layer for this purpose.

Figure 12-2. MOSFET Cross-Section

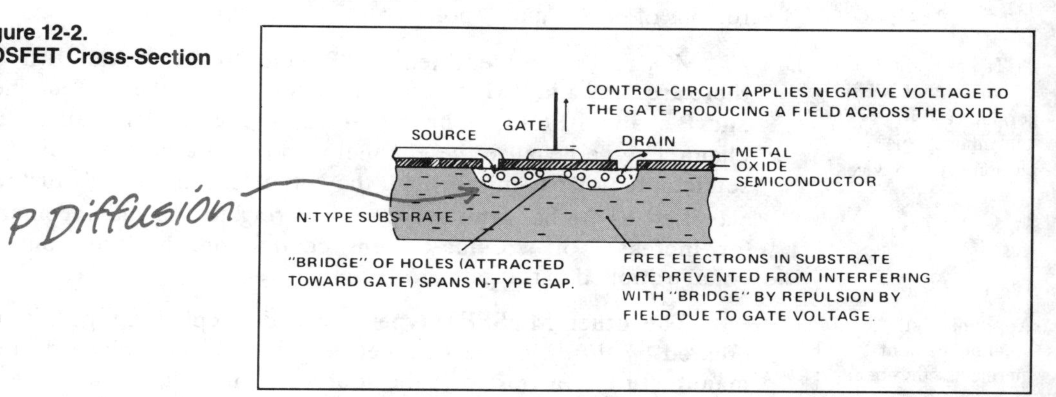

Applying a voltage to the gate creates an electric field from gate to substrate through the oxide layer. The field attracts current carriers to this region, forming a "channel" from source to drain for the working current.

Now suppose that we connect the gate terminal to a control circuit which applies a negative voltage to the gate. The gate is insulated, so control current cannot flow continuously. But a quick spurt does come out. The voltage on the gate produces a field across the oxide insulation from metal gate to N-type substrate. The field forces electrons in the N-type substrate *away* from the gate. At the same time the holes created are attracted to the gate. The insulating oxide prevents the holes from reaching the gate, so they distribute themselves right under the oxide, forming a bridge from source to drain through which working current can pass. This bridge is in effect a strip of P region, and is called the *channel*. The more negative the gate voltage, the thicker the channel becomes, and the more working current flows.

Bipolar transistors are controlled by current. Field-effect transistors are controlled by voltage.

So the MOSFET behaves much like a PNP transistor. But whereas the PNP transistor is turned on by *current* applied to the control terminal, the MOSFET is turned on by *voltage*.

The name "field-effect" applied to this type of transistor refers to the way it is turned on — by the *effect* of the electric *field* (a region of force or influence) created by the gate voltage. The term "metal-oxide-semiconductor" describes the *structure* in which this electric field is created — a metal gate, an oxide layer, and a semiconductor channel. (We will see in a moment that some FETs do not employ the MOS structure.)

Another name for FETs is "unipolar," meaning single polarity; this refers to the fact that working current passes through only *one* type of semiconductor material as it flows from source to drain — in our example, it is P type. "Bipolar" transistors are so called because their working current flows through regions of *both* P and N polarity.

FETs are manufactured in P-channel and N-channel versions as either "enhancement-type" or "depletion-type" devices.

MOSFETs are identified by "P" and "N," however. Just as there are NPN and PNP bipolar transistors, so are there N-channel and P-channel FETs. The MOSFET of Fig. 12.2 is called "P-channel" type because the channel is in effect a bridge of P-type semiconductor passing through the N-type substrate. It's further classified as "enhancement-type," referring to the enhancement (or increase) of working-current conduction, by the control voltage applied to the gate.

Applying a gate voltage to an enhancement-type FET increases current flow; applying a gate voltage to a depletion-type FET decreases current flow.

(The other MOSFET type is called "depletion-type"; it is created by *diffusing* the channel permanently into place during manufacture. In this way, the source, channel, and drain are a continuous strip of P-type or N-type material. The depletion-type MOSFET is normally turned on, and it is turned off only by applying a control voltage to the gate. A *positive* gate voltage must be applied to a P-channel depletion type to repel and choke off hole current, and a *negative* gate voltage must be applied to an N-channel depletion type, to repel free electrons and throttle their flow. As it turns out, however, the enhancement-type we have used for our example is far more useful in ICs.)

Figure 12.3 shows the symbol for the P-channel enhancement-type MOSFET of Fig. 12.2. The broken bar across the middle represents the channel, which is normally interrupted and requires a gate signal to *enhance* conduction. The arrowhead pointing from channel to substrate indicates a P-channel device;

Figure 12-3. P-Channel Enhancement-Type MOSFET

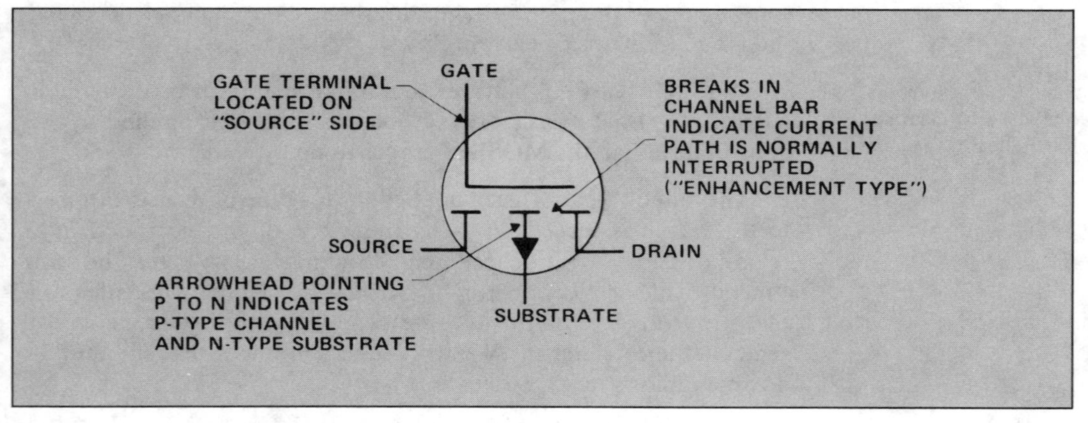

recall that arrowheads stand for junctions, and point from P to N. The source and drain are distinguished by the presence of the gate terminal over the *source*.

HOW DO MOSFETS PERMIT HIGH COMPLEXITY?

The small size and relative simplicity of MOSFETs permit integrated circuits of high complexity.

Many unique features of the MOSFET work together to permit integrated circuits of fantastic complexity:

First, a MOSFET takes up *much less room* on a chip than a bipolar transistor does, since only one diffusion is required instead of three. Figure 12.4 shows the relative sizes of the two types of

Figure 12-4. Comparative Size of MOS and Bipolar Transistors

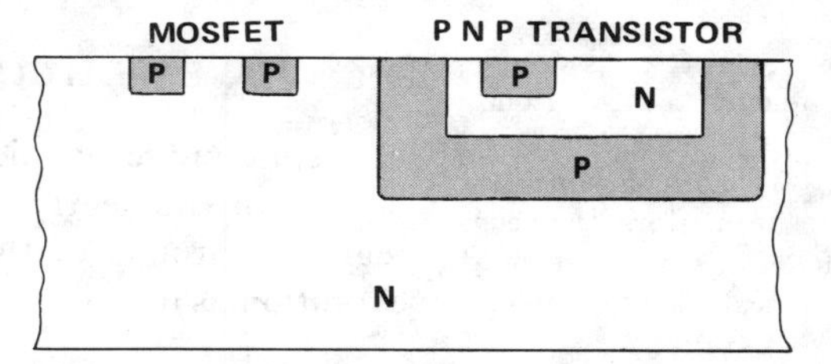

device. This single-diffusion process carries with it two additional blessings — processing is less costly, and the yield of good chips from each slice is higher, since diffusions tend to introduce defects.

Second, MOSFET circuits are usually *simpler* than equivalent bipolar transistor circuits. Figure 12.5 is a fairly typical example.

Figure 12-5. Relative Complexity of MOS and Bipolar NAND Gates

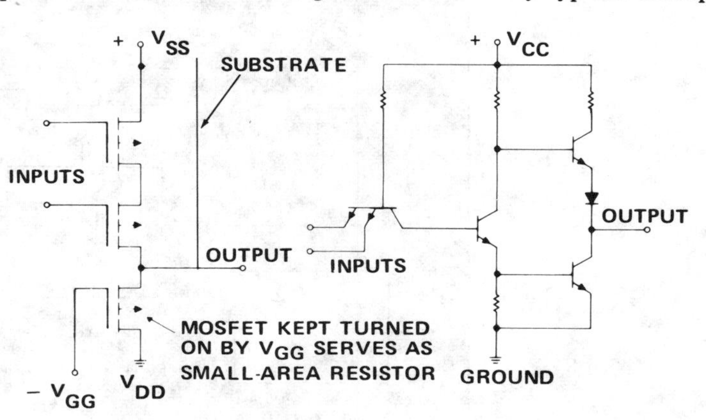

a. MOS Positive "Nand" Gate

b. 54/74 TTL Positive "Nand" Gate

Notice that the MOS gate has only three components, compared to nine for the bipolar gate. So not only are the components smaller, but fewer of them are required.

MOSFETs, used in circuits as resistors, can save silicon area.

Third, as Fig. 12.5 also shows, a MOSFET *can be used as a resistor,* by making its channel region somewhat longer and narrower than usual, and connecting its gate to a constant low-voltage supply (called "V_{GG}") so that it stays turned on. MOSFETs used in this way as resistors are hardly larger than normal MOSFETs. On the other hand, diffused resistors of the type used in bipolar ICs, if they are to have appreciable resistance, must be extremely long and occupy much more room than a bipolar transistor.

HOW HAVE HIGH COMPLEXITY DEVICES EVOLVED?

Refinements in semiconductor technology over the last 30 years have made first transistors, then complete integrated circuits practical and affordable.

Semiconductor digital devices have evolved from discrete devices in the early 1950's to the very complex large scale integrated circuits of the 1970's. We will now travel this road of evolution as it is summarized in Fig. 12.6.

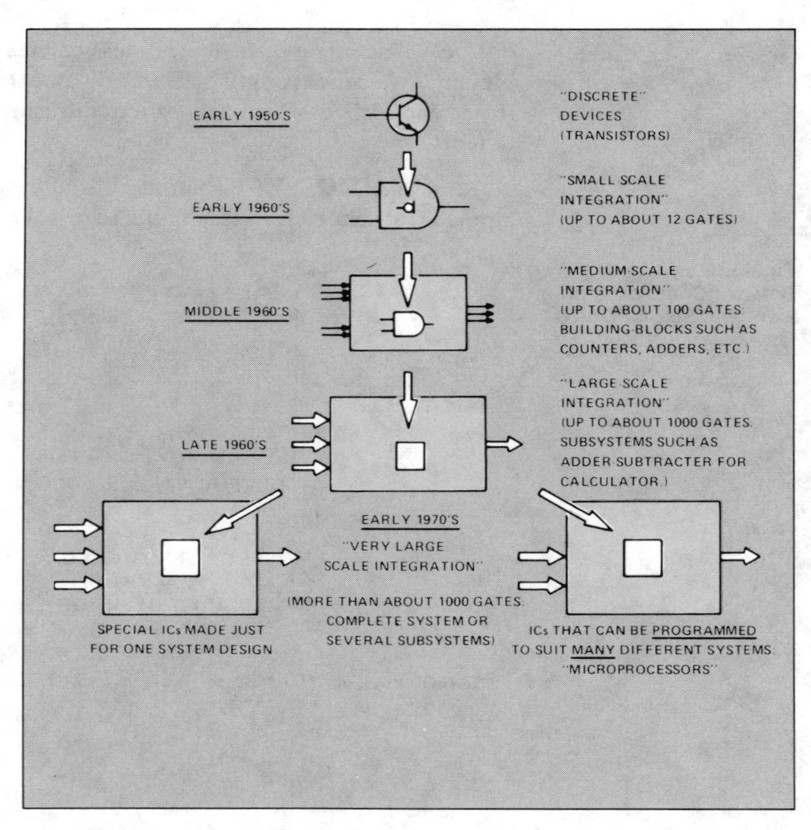

Figure 12-6. Evolution of Integrated Circuits
(G. McWhorter, Understanding Digital Electronics, *Texas Instruments Incorporated, Dallas, Texas, © 1978)*

In the mid-1950's, vacuum tubes dominate electronic circuits.

Let's begin our travels in the mid-1950's. Rock and roll is played on radios and phonographs built with bulky, hot vacuum tubes. Calculators are bulky. Computation is by slow slide rules or noisy electric adding machines. You see advertisements for transistors — discrete semiconductor devices of limited capability that cost a small fortune. When you build that hobby radio kit or design a photocell gadget, you use vacuum tubes. Maybe transistors are a passing expensive fad.

In the late 1950's, transistors are readily available and integrated circuits have just been invented.

Let's advance the clock to the late 50's. You are not much older, you still hear rock and roll, but transistors have become relatively inexpensive. There is a wide variety of types of transistors. Computers are switching over to transistors from vacuum tubes. Even small transistor radios are becoming available. Not only that but you hear about a device called an integrated circuit that's just been developed. The circuits consist of a few logic gates in a single small package. Well, the transistors are inexpensive and don't produce much heat, so you begin to use them in your hobby equipment or your engineering designs. But what's an integrated circuit? They cost a small fortune and there is only a limited selection available. So you continue to build your electronic equipment with resistors, transistors, diodes, and other components. It takes time, but it's not too expensive in terms of parts cost. The equipment works good. However, you would not be surprised if integrated circuits were here to stay, considering your experience with the transistor take-over.

In the mid-1960's, transistors have proven to be high-performance, low-cost, and reliable devices, and MSI digital IC's are moving into volume production.

Now it's the mid-sixties and rock and roll has been replaced by Beatle-mania. But integrated circuits are now everywhere. Computers are built with them. They are even appearing in television sets. What has happened? The maturing of semiconductor technology is what has happened. At first it was a struggle to make a working transistor. Most transistors made were defective and could not be sold. Within two or three years the processes used in manufacturing them improved so that most of the transistors were exceptionally good and very reliable. They required smaller area on the silicon slice and were becoming inexpensive. It became easy to start producing integrated circuits, especially digital integrated circuits. The computer industry was able to buy large numbers of these devices, making them very inexpensive. Why? It is a fundamental rule of the semiconductor industry that the larger the number of devices sold (large volume of sales and production), the lower the cost of the device. The processes improve rapidly with high volume, resulting in a high percentage of good devices (high yield) and highly reliable devices. Further, large initial design costs can be averaged out over millions of devices, so that even this cost is not important. So now you are flooded with IC's. They perform complete digital functions, such as counting, decoding, and storage. They are made with high performance TTL and ECL logic instead of the slower DTL and RTL logic. What do you do? Use them. It's far cheaper, easier, and takes less space to plug in a device someone else has built for you than it is to build your own. So you start using digital IC's even in areas that had been all analogue functions before. Your equipment costs less, weighs less, takes up less space, is cooler, lasts longer, and is easier to repair than ever before. Things are about as good as they can get. Right?

In the late 1960's, digital IC's contain entire subsystems, called LSI.

Wrong! It's the late 60's. What had been a fast paced evolution has become a revolution. IC's with 1000 gates per package are now available with the perfection of high quality small MOS devices (recall Fig. 12.4?). Now each package contains entire digital sub-systems such as memory and computational units. For the next decade, a continuing stream of new subsystems are offered including microprocessors (the central element of a computer) and even complete computers on a single integrated circuit made with MOS technology and each one seemingly more complex than the one before. Are they expensive? Initially, yes, but as more people buy them, the price comes down to levels lower than some single transistor devices sold for in the 50's. Are they difficult to use? No, someone is providing you with complete digital subsystems in the palm of your hand. You just connect the subsystems together to build systems that were totally beyond your capability or budget just a few years ago. All you have to learn is what these subsystems are and how to use them. The main subsystems available to us are memories, microprocessors, microcomputers, and input-output interface units.

In the 1970's and 1980's, IC's of all types have increased in density such that complete systems are contained on one chip with from 50,000 to 100,000 active devices per chip. VLSI (very large-scale integration) is the name.

The journey is complete and the time span is but a brief 30 years of semiconductor development. Through this period we have seen the most remarkable pattern repeating itself. New devices, more complex than previous devices (but easier to use) are introduced, usually at a high price and with limited variety. In a short period of time, the price drops abruptly, while the variety and quality of devices improves significantly. All along the continuing desire for greater reliability and lower system costs kept pushing the semiconductor industry to ever more complex devices. Let's now see how this pattern has occurred specifically for the evolution of complex MOS devices.

HOW HAVE COMPLEX MOS IC'S EVOLVED?

A need for small and fast computer memory chips, coupled with the success of electronic calculators built with MOS technology, resulted in the high complexity of present MOS integrated circuits.

It's easy to see how the technology could evolve from discrete transistors to bipolar integrated circuits. It's even easier to see how simple MOS integrated circuits developed. But what caused the push toward the complexity we have today in MOS devices? The first impetus arose from the computer manufacturer's need for small, fast memory. The second resulted from the ability of the semiconductor manufacturer to put all the functions of a four-function electronic calculator on several chips. It was an immediate hit and it was fed by the consumer's fascination and demand for them in the marketplace.

The MOS devices introduced in the late 60's found immediate application as high capacity memories, which could store up to 1000 binary bits of information per package. Many systems require such memories to save intermediate computation results, inputs, repeating bit patterns, and so on. Memories became a natural application area for MOS devices. Memories are made by repeating a basic storage cell circuit along rows and columns in a rectangular or square pattern as shown in Fig. 12.7. Since the basic MOS storage cell has such a simple geometry, the regular row and column pattern is easy to implement. The overall repetitive pattern can be generated automatically by computers and the design costs are low. Further, since a simple pattern is used, the number of good devices is increased so that cost of manufacture is low. As a result, MOS memories became economical and found wide application in computers and other systems requiring fast memories.

The MOS technology was a natural for the calculators. At first only simple four function calculators were possible. Even these required a level of integration that would not have been possible without high quality, small size MOS devices. Again, as techniques improved, scientific and business function and even programmable

Figure 12-7.
MOS Memory Cells

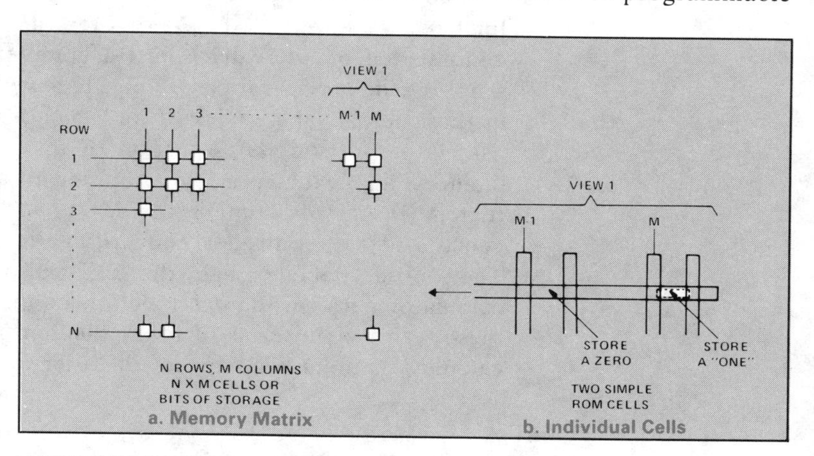

a. Memory Matrix
b. Individual Cells

Figure 12-8. Calculators Made Possible by MOS Technology

calculators emerged, as shown in Fig. 12.8. Initially, calculators required two large scale integrated (LSI) circuits. One circuit provided the memory for the calculator and the other provided the computational capability. This evolved into a single LSI chip calculator. The result of the single chip approach was lower costs even for calculators with increased functional capabilities.

Initially, calculators needed one IC for memory and another for computation. Present models implement both functions on a single chip.

To appreciate the accomplishment of the calculator, consider the calculator chip and block diagram shown in Fig. 12.1. This chip requires a second chip to provide the instruction sequence (program) memory for the calculator. All other functions are provided on the single chip of Fig. 12.1. There is an instruction register to store the current instruction and an instruction decoder to determine which instruction is to be executed. This information is used by the control decoder and associated timing and counting circuitry to cause the system to perform the proper sequence of operations. The basic arithmetic and logical operations are performed by the arithmetic logic unit (ALU) with intermediate and final results stored in the 24 register file. All of this may seem complicated but it really is just a complex version of our basic sense-decide-act operations. In Fig. 12.9 we consider the calculator in these terms. The keyboard and encoder part of the complete calculator are the *sense* portion of the system. The instruction memory and calculator chip register file memory perform the remember or *store* function. The *decide* portion of the system consists of the controller and arithmetic logic operational unit. The output decoder and the LED display of the completed calculator perform the *act* function of the system. Looked at in these terms or in terms of block diagram functions, the complex calculator system becomes much easier to understand. The single

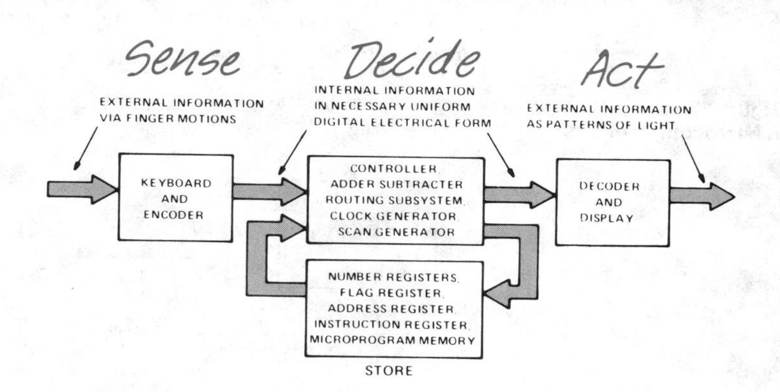

Figure 12-9. Block Diagram of a Calculator
(G. McWhorter, Understanding Digital Electronics, *Texas Instruments Incorporated, Dallas, Texas © 1978)*

chip calculator, like the one of Fig. 12.1, does all of the operations we have looked at, and provides all of the capability, including program memory.

With the complexities represented by calculators and large memories in a single LSI circuit, the potential for placing other entire digital systems in a single integrated circuit began to be realized. Such systems could be designed for some special task such as an automotive controller or a specialized control center for a manufacturing process or an appliance control, or similar systems. Eventually, the digital system could be a complete general purpose computer in a single IC. These are the options available to us today in digital integrated electronics. In order to understand what's involved in this choice, let's examine what is meant by microprocessor and microcomputer systems.

WHAT ARE MICROPROCESSORS AND MICROCOMPUTERS?

Microprocessors are integrated circuits that control the overall functions of all types of digital systems.

The experience gained in providing MOS memory and calculator chips provided the background for developing the microprocessor on a chip followed by the computer on a chip. All of us know what the calculator does and what it looks like, though we may not know how it works. A microcomputer is like a calculator but it can do more. The basic building blocks of a microcomputer are shown in Figure 12.10, many of which are common to calculators. The microprocessor is the center of action. It is a single IC that controls the operation of the entire system. It can follow instructions to do arithmetic, logic, compare, make decisions. The complete system is more than a calculator because it takes inputs from various type units and directs signals to various outputs. It also usually has more memory. Initially, each memory and input-output block consisted of one or more separate integrated circuits. In order to understand the operation of this system, we must understand the purpose of its building block components.

Figure 12-10.
A Microcomputer System

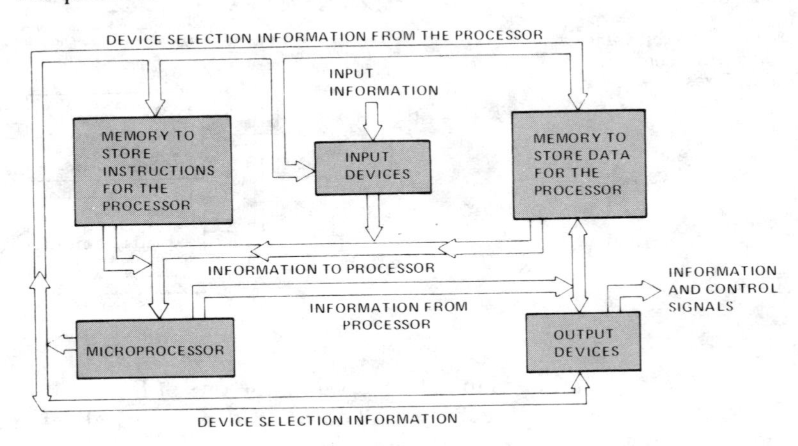

Microcomputers, small microprocessor-controlled systems, more elaborate than calculators and with more memory and input/output capability, are programmable using computer languages.

The instruction storage memory stores the sequence of instructions (called a program) that the processor will use to decide what to do. The program then defines what operations the system performs and how it behaves.

All microcomputer systems have this same basic structure of Fig. 12.10. It is primarily the program that distinguishes one system from another. One program may cause the system to act as a controller for a manufacturing operation; another program makes the system behave as a calculator; another program can cause the system to act as a switchboard in a communications or information network. Thus, to change the system operation, we mainly just change the program in the instruction memory. Since the processor just reads the information from the instruction memory, it is called read-only memory, abbreviated as ROM.

The data storage memory contains binary data that is needed by the processor as it executes the instructions. This data may be information that has been input into the system just waiting to be used by the processor and its program. It may be information that has just been calculated and is to be sent out of the system by the processor. It may be results of processor operations that need to be saved for future use. This memory is often called random access memory (RAM for short) or read-write memory since the processor may send data to any memory location (a write operation) or it may get data from any memory location (a read operation).

The input devices allow the external subsystems to send information to the system of Fig. 12.10. These may be keyboards like on a calculator or typewriter, or they may be telephone inputs, or special sense circuits for tape machines. The output devices allow the system to send information out to display and control elements. The display element could be LED displays, like on watches or calculators, printers, or a television screen. The control elements could be power semiconductors, relays, motor controls, and so on.

A microprocessor handles its program as a list of instructions, which it interprets and executes one at a time in sequence.

The microprocessor is the central decision and control element for the system. Its job is to obtain the instructions from memory (ROM) and execute them. The processor acts much as we do in following instructions. For example, if at the beginning of a day we receive a list of jobs that have to be done that day in a certain sequence, we will examine that list starting at the first job, as shown in Fig. 12.11. We determine what the first job requires of us, get together all the things we need to do that job, and then complete

Figure 12-11. Program-Job List Analogy

the job. We go on to the next job on the list, and repeat the procedure until all jobs have been completed. We are acting just as the processor in Fig. 12.10 does. We get the instruction, which may be mow the lawn. We then get all the equipment we need to do the job (mower, grass bags, etc.) We then mow the lawn, put up the equipment, and then go on to the next job on the list. The list of jobs to be done is like the ROM program in the microcomputer. The equipment needed to do the job is like the data in RAM or the input data or information from output devices. The processor is like the person doing the work. The main difference between the human processor and the microprocessor is the instruction capability of the two systems. Humans can interpret and execute very complex instructions. Microprocessor instructions are very elementary. They provide for moving data around in the system, performing simple arithmetic and logical operations on this data, and making basic yes-no decisions.

Their programmability makes microcomputer systems extremely versatile: by changing the program, you can prepare a system for an entirely new application.

The processor along with ROM make up the *decide* portion of our system. The input devices and RAM make up the *sense* and *store* portion of the system. The output devices are the *act* portion of the system. Since the system contains all the required elements to perform any function, we have a universal system structure in the microcomputer. By using the appropriate input devices, output devices, and program, the system structure can be used to perform any function we desire. Further, by just changing the program and maybe an input or output device, a relatively simple process, we can completely change the application and function of the system.

The microcomputer has many features that make it an ideal approach to digital systems design. Each system consists of only a few of the same integrated circuits. Since many people are using these same devices for their systems, the cost per device is very low considering the complexity and power of these LSI devices.

Recent developments have made it possible to put a complete microcomputer on a single chip.

As attractive as the microcomputer structure of Fig. 12.10 is, there is one further step in the evolution of digital LSI elements that makes it even more attractive. This step is the production of all the elements of Fig. 12.10 in a single IC — a microcomputer in a single package! With this development, it is possible to provide entire digital systems or subsystems on a single small integrated circuit. We only need to attach our sense or input devices, power supplies, and output display and control devices to complete the system. By building or programming the microcomputer with the appropriate program, the system will behave as required by the application. We only buy one relatively low cost universal digital system element and add on the accessories and program to do what we want.

Figure 12-12. Microcomputer

a. Discrete Component Version

b. Single Chip Version

To gain an appreciation of the size reduction this provides for us, consider the alternatives in Fig. 12.12. Fig. 12.12a shows a microcomputer on a board using individual IC's. Fig. 12.12b shows the same functional capability as a single IC!

HOW ARE COMPLEX MOS DEVICES BEST USED?

Digital systems for specific applications may be built either with specially designed chips or with general-purpose microcomputers running specially developed programs. In most cases, the microcomputer approach is the more practical.

We have two distinct approaches that can be used to apply LSI circuits to building digital systems. We can use special purpose designs that work in only one specific application, or we can use a general purpose microcomputer and apply it to each system application we have by changing programs.

Let's say we design a special purpose LSI circuit that will perform the specific system task we want. This was done in the case of the electronic calculators. It could be done for special purpose automobile control computers. The advantage to this approach is that the design can be such that it interfaces more or less directly with the system it is used in, so that only basic input and output devices have to be purchased. The disadvantage is that unless very large numbers of such special LSI circuits are purchased, the cost per device is much too high to be considered. Thus, this approach can be used only in applications that involve a high number of identical units, such as in the case with calculators and automobile systems.

In contrast, a microcomputer can be used to make any digital system by using the appropriate program and input-output components. Further, if many different users adopt this approach, the cost to each user of the system components will be low enough to be practical. For people who build only a few of a given type of system, the microcomputer approach is a very economical approach and the special purpose LSI device approach is not possible. Even for people who build a large number of systems, the microcomputer on a chip approach may be better than the special purpose LSI circuit approach. To adapt the microcomputer to a new but slightly different problem (say a new automobile) involves only minor program changes. To adapt the special purpose LSI device to the new application would require an entire and very expensive redesign of the LSI device. Thus, only in cases where the system requirements are fixed and there are a to be millions of identical systems made is the special purpose LSI approach sensible. All other applications would best be met with the programmable microcomputer approach.

WHERE ARE DISCRETE FIELD-EFFECT TRANSISTORS USED?

Discrete FETs are important in applications where voltage, not current, is the controlling parameter. One such application is the amplification of a low-current, high-voltage signal from a high-impedance microphone.

As long as we are talking about field-effect transistors, we should digress a moment and mention the use of discrete FETs. Although tiny transistors, they are important in some applications where we need a transistor that is controlled by *voltage*, rather than by *current* as bipolar transistors are.

Figure 12.13 suggests a case in which voltage-controlled transistors are desirable. In this case, we have used *junction*-FETs, which are not made by the MOS process, but function very much like depletion-type MOSFETs. The problem here is to amplify the signal from a high-impedance microphone to drive the loudspeaker. A high-impedance microphone cannot generate very much *current;* instead, it generates relatively high *voltage*. With a high-impedance microphone, we cannot very well use a *bipolar* transistor amplifier as we did in Chapter 8, because bipolar transistors are operated by control current. It would take several bipolar amplifier stages in a row to amplify the small current we get from a high-impedance microphone.

Figure 12-13. High-Impedance Microphone Amplifier

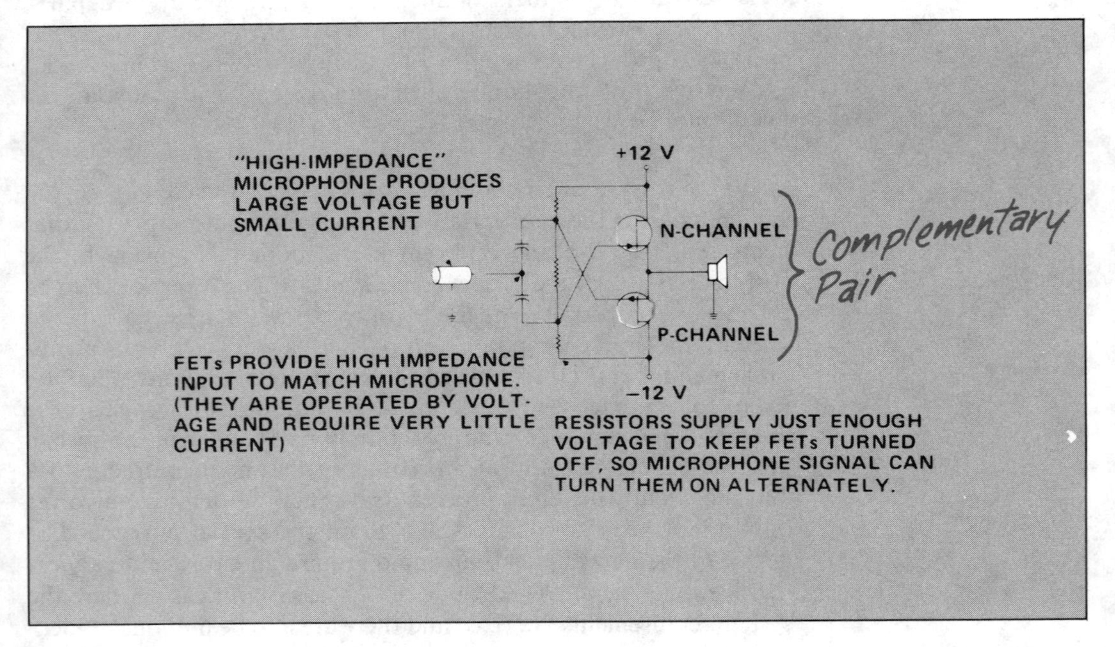

Instead, we use FETs as in Fig. 12.13. They require very little control current. The voltage fluctuations from the microphone let the FETs produce current fluctuations of sufficient power to drive the loudspeaker. We can say that "FETs have a high-impedance input." This characteristic matches our microphone, and the problem is solved.

FETs find many other applications, and are used most often because they control working current in response to *voltage* rather than current signals.

Quiz for Chapter 12

1. The main advantage of using MOSFET rather than bipolar transistor circuitry in ICs is:

- ☐ a. Higher operating speed than bipolar circuits
- ☐ b. Fewer power supply connections are required with MOS ICs
- ☐ c. Much greater complexity (more gates and flip-flops in less chip area) than bipolar circuits—therefore better economy
- ☐ d. System designers are more familiar with MOS circuitry
- ☐ e. There is no particular advantage

2. The main advantage spoken of in question 1 is due to the fact that:

- ☐ a. MOSFETs occupy much less chip area than bipolar transistors
- ☐ b. Several separate layers of circuitry are stacked up in MOS chips
- ☐ c. MOSFETs can be used as resistors and are much smaller
- ☐ d. Digital MOSFET circuitry is generally quite simple (few components per gate)
- ☐ e. All but b above

3. A P-channel enhancement type MOSFET (the sort used in MOS ICs) performs much the same function as a PNP transistor, except that:

- ☐ a. It is considerably larger
- ☐ b. It operates much faster
- ☐ c. It regulates but does not switch
- ☐ d. It is controlled by voltage rather than by current, so that it requires very little current at the control terminal (gate)
- ☐ e. It is controlled by current rather than by voltage like a bipolar transistor

4. MOS IC chips are easy and economical to process because:

- ☐ a. They are very low in complexity
- ☐ b. The chips are much smaller than for bipolar transistors
- ☐ c. Only one diffusion step is required, compared to three for bipolar ICs
- ☐ d. They do not require any diffusion at all
- ☐ e. a and b above

5. The terminals of a field-effect transistor (whether MOS or junction-type, enhancement or depletion type), corresponding respectively to the emitter, base, and collector of bipolar transistors, are

- ☐ a. Source, gate, and drain
- ☐ b. Anode, gate, and cathode
- ☐ c. Input, power supply, and output
- ☐ d. Metal, oxide, and semiconductor
- ☐ e. Drain, channel, and source

6. The evolution of semiconductors has resulted in:

- ☐ a. Higher performance IC's
- ☐ b. More expensive systems
- ☐ c. Increased reliability
- ☐ d. More digital devices per IC
- ☐ e. a, b, and d above
- ☐ f. a, c, and d above

7. What devices pushed the MOS technology to their present high levels of complexity?

- ☐ a. Dimmer controls
- ☐ b. Calculators
- ☐ c. Memories
- ☐ d. Printers
- ☐ e. None of the above
- ☐ f. b and c above

8. The best way to utilize MOS LSI devices in systems is:

- ☐ a. Limited copies of special purpose LSI devices
- ☐ b. The microcomputer LSI chip
- ☐ c. The microprocessor-microcomputer structure
- ☐ d. Program controlled digital system structures
- ☐ e. All but a above

9. Program Controlled digital system design:

- ☐ a. Is not versatile
- ☐ b. Is hard to design
- ☐ c. Offers simple hardward design
- ☐ d. Is widely applicable to all systems
- ☐ e. c and d above
- ☐ f. none of the above

13 LINEAR INTEGRATED CIRCUITS

Key Words

Comparator
Darlington Configuration
Decibels
Frequency Response
Logarithm
Op-amp
Systems Interface Circuits
Voltage Gain

Definitions are found in the glossary in the back of the book.

Linear Integrated Circuits

WHAT ARE LINEAR INTEGRATED CIRCUITS?

Linear integrated circuits are amplifying-type circuits in integrated form. The name "linear" is just another way of expressing the concept of "regulating" as opposed to "switching." The output of such a circuit changes in a smooth, even manner as the input is changed at a constant rate, so that a graph of output versus input is approximately a straight line — hence the name *linear*. In contrast, as you have seen, the output of a switching circuit jumps suddenly from one level to another.

Linear ICs (amplifying circuits in integrated form) can be grouped into three categories: general linear circuits, systems interface circuits, and consumer-and-communications circuits.

Since linear circuits are amplifiers, their function is to *increase* the power, current, or voltage applied to their inputs.

WHAT ARE SOME REPRESENTATIVE LINEAR ICs?

Since you now have a clear idea of how discrete transistor circuits perform their regulating functions — their linear functions — we can discuss linear ICs without going into their internal circuitry. And since there are thousands of potential linear IC types, we must cover the ground by talking about only a few *representative* types.

Linear ICs can be classified roughly into three groups: general linear circuits, systems interface circuits, and consumer-and-communications circuits.

The most versatile of the general linear circuits are operational amplifiers, originally developed for analog mathematical operations.

General linear circuits perform the amplifying function in a great variety of systems. Included in this circuit category are operational amplifiers, video amplifiers, voltage comparators, voltage regulators, and several other types of amplifiers. These were the first linear circuits to be integrated successfully.

Operational amplifiers, informally called "op-amps," are probably the most important circuits among all linear ICs. An op-amp has extremely high voltage gain, and differential inputs (recall we discussed differential inputs in Chapters 2 and 3). Op-amps were originally used to perform mathematical operations in analog computers, but now they are used wherever low-power amplification is needed, in almost any system. They can be used as the principal amplifying element in a wide variety of applications, simply by adding a few components; applications range from controllers for TV antenna rotators, to radio transmitters.

Systems interface circuits amplify signals entering or leaving digital systems.

Systems interface circuits essentially amplify information signals going into or coming out of digital systems. They act as the "go-betweens," or *interfaces* that permit the various sub-systems of a complete digital system to be coupled together. At this writing, interface circuits can be classified into three major groups:

Memory drivers and *sense amplifiers* serve as writing and reading units, respectively, for magnetic memories. The drivers get digital information into the memories, and the sense amplifiers get the information out.

Line drivers and *line receivers* perform much the same function, but at the ends of long transmission lines between parts of a digital system. They transmit and receive digital information—sometimes across a continent.

Peripheral drivers are quite similar to memory drivers and line drivers. But they drive the peripheral equipment of a digital system — the output units, such as printers and information displays. Peripheral drivers accomplish this by translating the low-power digital signals generated by the *"decide"* stage, into the high-power signals required by the "act" stage.

Consumer-and-communications circuits are used in home and automobile electronics and in military and industrial communication systems.

Consumer-and-communications circuits tend to be overlapping categories. Consumer circuits are ICs used in home-entertainment equipment and automobile electronics. Included are a wide range of amplifiers, from audio (sound) amplifiers to drive small loudspeakers, to radio-frequency signal processors. Communications circuits are a group of amplifiers used in military and industrial radio systems.

WHAT IS REQUIRED OF LINEAR IC TECHNOLOGY?

The basic planar process we discussed in chapter 7 is used for both linear and digital integrated circuit fabrication. However, these different types of devices require different types of performance. The digital circuits require high switching speed and

The performance required of linear ICs imposes some special steps during their manufacture.

high conductivity devices. The linear circuits require high gain and high frequency of operation. Further, it is usually desirable for linear circuits to draw a minimum of current from the circuits to which they are interconnected. That is, they should have a high input resistance. To achieve these features of high gain, high input resistance, and high frequency response requires special circuit and fabrication techniques which we'll look at in this section.

HOW IS HIGH GAIN ACHIEVED?

Reducing base current by narrowing the base region increases current gain. It also improves frequency response.

To achieve high gain, we need transistors with large current gains (I_c/I_b) and large resistor values. The most common method of increasing current gain is to decrease the width of the base region. The effect of reducing the base width is illustrated in Fig. 13.1. If both transistors have the same electron flow through the N regions from emitter to collector, both bases will have the same number of electrons flowing through them per second. They pass through the narrow base region in half the time it takes them to pass through the wide base region device. This extra time in the wider base causes more opportunities to fill holes in the base, causing the base current to increase. The narrow base device has less filling of holes by the electrons and thus less base current for the same collector current. Very narrow base devices are called super-beta transistors because of their very high current gain. The main problem with such devices is that they cannot provide a very large output voltage without drawing excessive short circuit currents from the collector to the emitter. In other words they have a low collector to emitter breakdown voltage.

Figure 13-1. Comparison of Conventional and High-Gain Transistors

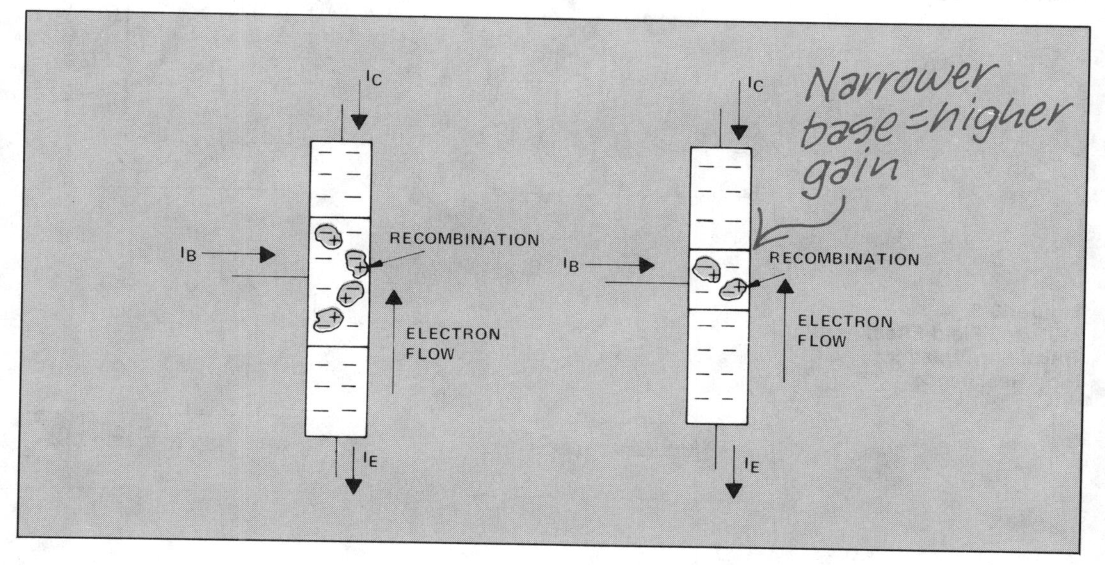

HOW IS HIGH RESISTANCE ACHIEVED?

High resistances can be created where necessary with active devices such as PNP or MOS transistors.

To achieve high resistance values without using diffused resistors which require a large amount of chip area, linear circuits use resistors formed with active devices similar to the way shown in the MOS circuit of Fig. 12.5. Two common ways of making resistors (really active devices) for linear circuits use either a PNP transistor or a type of field effect transistor. PNP transistors are difficult to fabricate with the planar process that is set up for fabricating NPN transistors. Even so, low performance PNP transistors can be made and have been used as shown in Fig. 13.2. A constant current is drawn from the base of the PNP transistor. This produces an approximate constant collector current over a large range of collector-to-emitter voltages. For collector currents within the linear range, the collector-to-emitter circuit looks like a very high resistance. Used like this, the PNP transistors provide large resistance values and thus a large voltage change divided by the input voltage change that caused it—V_o/V_in.

MOS transistors can be used in the same way as the PNP to achieve large resistance for large voltage gain. We can also use a FET (field-effect transistor) type of structure using diffusion to provide the narrow channel required to provide the desired high resistance. Such structures are illustrated in Fig. 13.3.

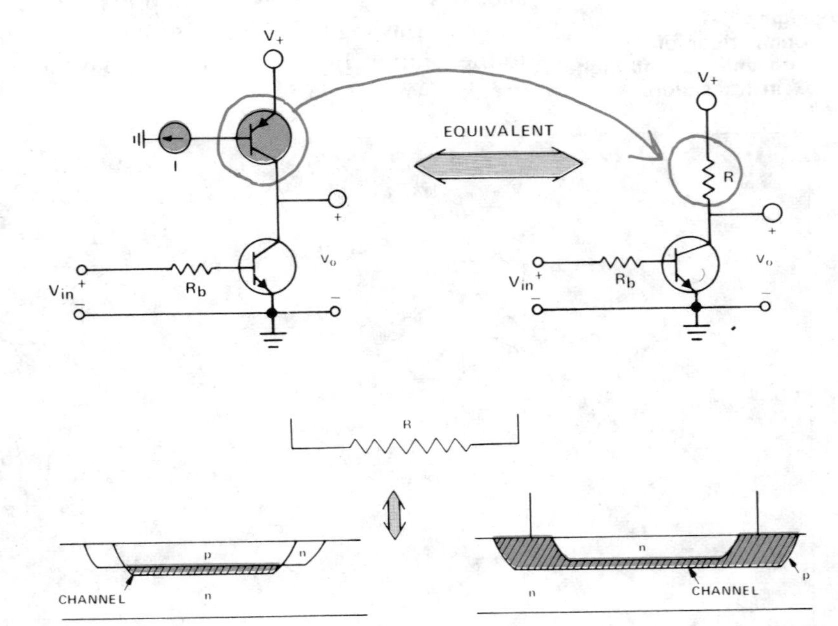

Figure 13-2. PNP Transistors Used to Provide High Resistance in Linear Circuits

Figure 13-3. MOS and Field-Effect Transistors Used for High Resistance

HOW IS A CONSTANT CURRENT ELEMENT USED?

The constant current produced by using active devices to create high resistances helps linear ICs maintain operating stability.

The same approach used to achieve high resistance for increasing circuit gain can be used to provide constant current sources required by linear IC's. Such constant currents are used in differential amplifiers to improve their performance. A constant current can be made using a large voltage and a large resistor. This is shown in Fig. 13.4a. If V is large and R is large compared to Rx, then Ix will change very little even though Rx is changed. Alternatively, we can bias a transistor to a certain constant current flow through its collector. The simplest circuit that does this is shown in Fig. 13.4b. Since both transistors are identical and they both have the same base to emitter voltage, they will both have the same collector current. We can control I_1 and thus I by the value of R. Then I will be held constant at I_1 as desired.

WHAT ARE OTHER LINEAR IC CIRCUIT TECHNIQUES?

Designing circuits with high input resistance and making gain dependent on resistance ratios rather than resistance values allows manufacturers to produce linear ICs with consistent and predictable characteristics.

The characteristics of linear circuits often depend on resistances being very close to some specified value. Integrated circuit processes produce widely variable resistor values. In order to use these IC resistors in linear IC's, the circuits are designed so that circuit performance depends on resistor ratios rather than resistor values. These ratios can be held relatively close to the desired value and as a result, the device characteristics tend to be relatively predictable from device to device.

Figure 13-4. Constant Current Sources

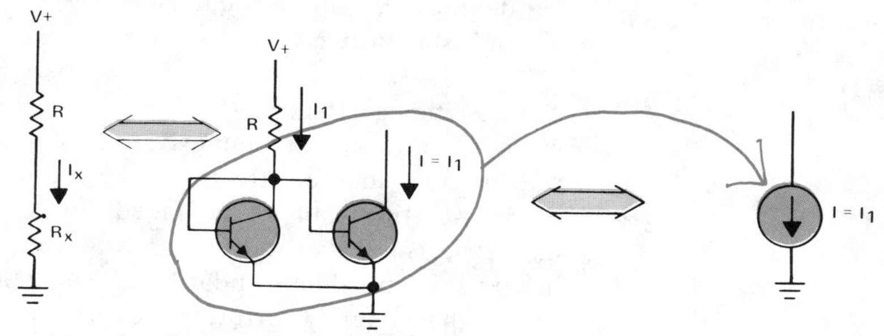

a. Large Voltage and Large Resistance **b. Transistor Biased for Constant Current** **c. Symbol**

When we produce transistors with high gain, we can reduce input base current and thus increase input resistance. One obvious way to provide an almost open circuit is to use a MOS transistor as the input device. It has very high input resistance. However, special processing is required to have the right MOS on a bipolar IC. Another approach is to use the Darlington configuration (recall Fig. 2.7?) to reduce input current. This configuration is shown in Fig. 13.5. If the circuit requires a certain collector current, I_C, then the base current required, I_{B1}, is I_C divided by the transistor's gain,

β. Putting a second transistor on as in the Darlington connection, causes the input current I_{B1} to be divided by β again. Thus, the input current is the output collector current twice divided by the transistor current gain β. This results in a very low value of input current and thus a high value of input resistance.

Figure 13-5. Darlington Transistor Provides High Input Resistance

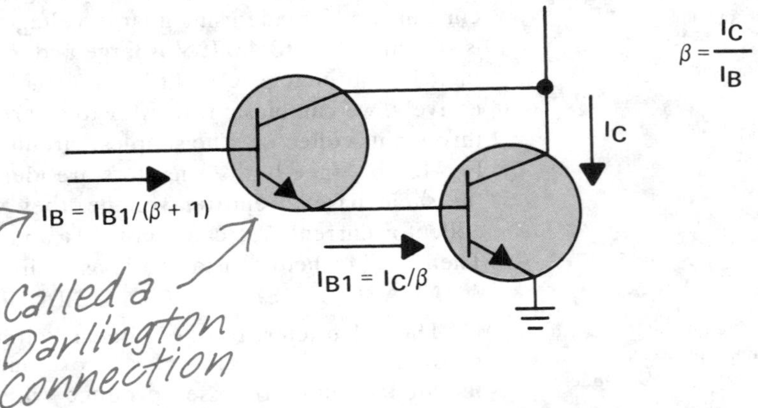

HOW ARE HIGH FREQUENCY DEVICES ACHIEVED?

By decreasing circuit area and increasing the isolation between components and ground, manufacturers can produce linear ICs for high-frequency applications.

High frequency devices must be of small area to reduce reactance effects within the device. To achieve really high frequencies, the reactance between components and ground must be decreased. The normal isolation between components and ground is a reverse biased silicon diode as illustrated in Fig. 13.6.

The reactance effect of these can be reduced by area reduction, but not to the extent to achieve really high frequency operation. To reduce the reactance effects between components even further, the diode isolation can be replaced by a true insulator called a dielectric. Such a structure is shown in Fig. 13.7. Such a process is more expensive than standard diode isolation, but it can be used if very high frequency performance is required.

Figure 13-6. Reverse-Biased Diodes for Isolation Between Components and Ground

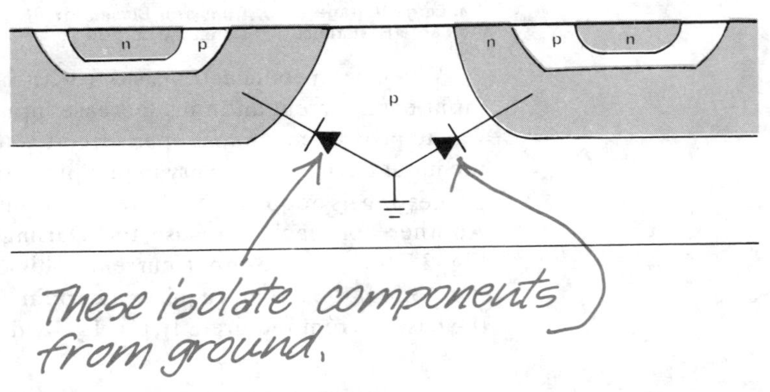

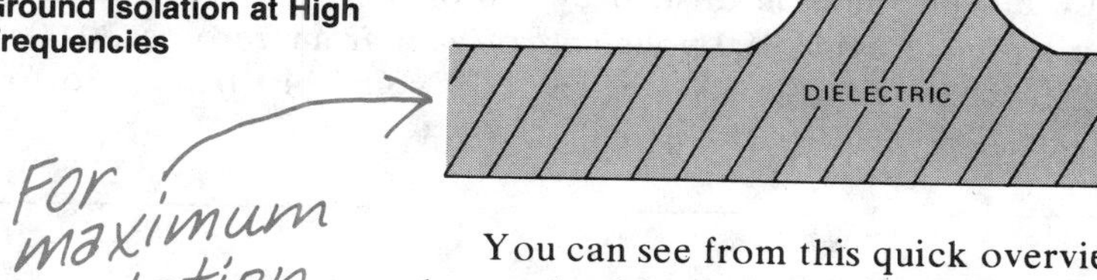

Figure 13-7. Dielectric Insulator for Improved Component-to-Ground Isolation at High Frequencies

You can see from this quick overview that linear IC technology has evolved from the same basic processes used for digital IC's. By varying these processes to best suit linear IC requirements, the economics of digital IC's can be applied to the linear world. This enables us to buy linear circuit functional blocks instead of tediously building them ourselves. These can be easily interconnected to build our linear subsystems and systems, with minimum design and construction efforts and costs.

WHAT DO LINEAR IC DATA SHEETS TELL US?

Parameters of interest on linear IC data sheets include gain, frequency response, input resistance, and power dissipation.

Since linear devices are basically different in their purpose from the digital IC, we would expect different parameters to be emphasized on linear IC data sheets. Figures 13.8 and 13.9 present a complete integrated circuit data sheet for an operational amplifier. This happens to be a combination field-effect transistor and bipolar transistor integrated circuit to give high input resistance. Of course, each different type of linear circuit will require different specifications. However, from our discussion of linear IC technology, we can see that the basic linear IC electrical specifications should include:

Gain — Since the linear device is often primarily an amplifier we would like to know the ratio of the change in output voltage to the change in input voltage. This can be expressed directly as a numerical ratio either in volts per volt or expressed logarithmically in units called decibels. $\text{Gain}_{db} = 20 \log_{10} \Delta V_o/\Delta V_{in}$. A gain of 100 volts/volt is 40 decibels or db, 1000 volts/volts is 60 db, and so on. The device TL081C of Fig. 13.8 and 13.9 has a typical gain A_{VD} of 200 volts/millivolt which is 200,000 volts/volt. In decibels this represents a gain of 106 db.

Frequency Response — This parameter is given as a curve or as a number on the data sheet. The TL081C data sheet (Fig. 13.9) indicates a unity gain bandwidth of 3 MHz typically. This means that the device has gains of one or more for frequencies up to 3 million hertz (cycles per second). This number can be used to find the bandwidth at any gain by dividing the desired gain into the 3 MHz number. For example, a gain of 100 volts/volt (40 db) would have a band-width of 3 million divided by 100 or 30,000 hertz. The amplifier would provide 100 volts/volt of gain from zero to 30,000 hertz. It would provide 20db of gain (10 volts/volt) from zero to 300,000 hertz.

Figure 13-8. Operational Amplifier Data Sheet

TYPES TL081AC AND TL081C JFET-INPUT OPERATIONAL AMPLIFIERS

electrical characteristics, $V_{CC\pm} = \pm 15$ V, $T_A = 25°C$ (unless otherwise noted)

PARAMETER		TEST CONDITIONS†		TL081AC			TL081C			UNIT
				MIN	TYP	MAX	MIN	TYP	MAX	
V_{IO}	Input offset voltage	$R_S = 50\ \Omega$				6		10	15	mV
		$R_S = 50\ \Omega$,	T_A = full range			7.5		13	20	
α_{VIO}	Temperature coefficient of input offset voltage	$R_S = 50\ \Omega$,	T_A = full range		10			10		μV/°C
I_{IO}	Input offset current				0.2	0.5		0.2	0.5	nA
		T_A = full range			0.4	1		0.4	1	
I_{IB}	Input bias current				2	4		2	4	nA
		T_A = full range			3	6		3	6	
V_{ICR}	Common-mode input voltage range			+12 to −12			+10 to −10			V
V_{OPP}	Maximum peak-to-peak output voltage swing	T_A = full range	$R_L \geq 10$ kΩ	24	26		24	26		V
			$R_L \geq 2$ kΩ	20			20			
A_{VD}	Large-signal differential voltage amplification	$R_L = 10$ kΩ, $V_O = \pm 10$ V	$T_A = 25°C$	25	200		25	200		V/mV
			T_A = full range	15			15			
r_i	Input resistance				10^{9}			10^{9}		Ω
CMRR	Common-mode rejection ratio	$R_S = 10$ kΩ		70	90		70	90		dB
k_{SVR}*	Supply voltage rejection ratio			70	80		70	80		dB
I_{CC}	Supply current	No load, No signal	$T_A = 25°C$		2	4		2	4	mA
			T_A = full range		3	6		3	6	

*$k_{SVR} = \Delta V_{CC\pm}/\Delta V_{IO}$.
†All characteristics are specified under open-loop operation, unless otherwise noted. Full range for T_A is 0°C to 70°C.

operating characteristics, $V_{CC} = \pm 15$ V, $T_A = 25°C$

PARAMETER		TEST CONDITIONS		TL081AC			TL081C			UNIT
				MIN	TYP	MAX	MIN	TYP	MAX	
SR	Slew rate at unity gain	$V_I = 10$ V, $C_L = 100$ pF	$R_L = 2$ kΩ,		9			9		V/μs

**Figure 13-9.
Operational Amplifier
Data Sheet (Continued)**

TYPES TL081AC AND TL081C
JFET-INPUT OPERATIONAL AMPLIFIERS

JUNE 1976

- JFET Input Stage
- High Input Impedance. . . 10^9 Ω Typ
- High Slew Rate Typically 9 V/μs
- Low Input Bias Current . . . 2 nA Typ
- Low Input Offset Current . . . 0.2 nA Typ
- No Frequency Compensation Required
- Continuous-Short-Circuit Protection
- Unity Gain Bandwidth . . . 3 MHz Typ
- No Latch-Up
- Low Power Consumption

description

This monolithic JFET-input operational amplifier incorporates well-matched, high-voltage BI-FET technology (JFET's on the same chip with standard bipolar transistors). The device features low input bias and offset currents, low offset voltage and offset voltage temperature coefficient, coupled with offset adjustment that does not degrade temperature coefficient or common-mode rejection.

The TL081C is characterized for operation from 0°C to 70°C.

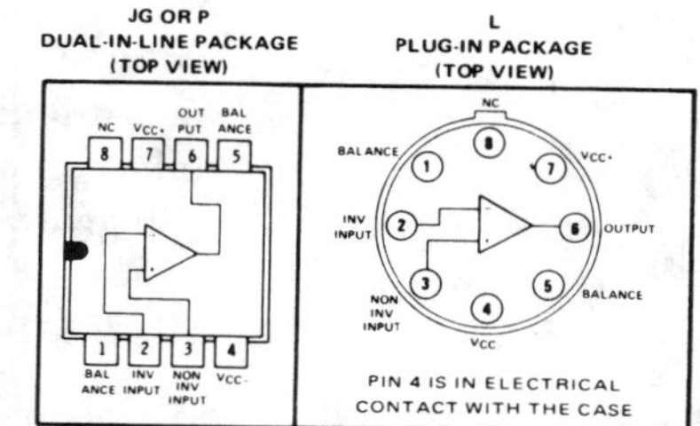

absolute maximum ratings over operating free-air temperature range (unless otherwise noted)

Supply voltage V_{CC+} (see Note 1)	18 V
Supply voltage V_{CC-} (see Note 1)	−18 V
Differential input voltage (see Note 2)	±30 V
Input voltage (see Notes 1 and 3)	±15 V
Duration of output short-circuit (see Note 4)	unlimited
Continuous total dissipation at (or below) 25°C free-air temperature (see Note 5)	670 mW
Operating free-air temperature range	0°C to 70°C
Storage temperature range	−65°C to 150°C
Lead temperature 1/16 inch from case for 60 seconds: JG or L package	300°C
Lead temperature 1/16 inch from case for 10 seconds: P package	260°C

NOTES:
1. All voltage values, except differential voltages, are with respect to the zero reference level (ground) of the supply voltages where the zero reference level is the midpoint between V_{CC+} and V_{CC-}.
2. Differential voltages are at the noninverting input terminal with respect to the inverting input terminal.
3. The magnitude of the input voltage must never exceed the magnitude of the supply voltage or 15 volts, whichever is less.
4. The output may be shorted to ground or either supply. Temperature and/or supply voltages must be limited to ensure that the dissipation rating is not exceeded.
5. For operation above 25°C free-air temperature, refer to Dissipation Derating Curves, Section 2. This rating for the L package requires a heat sink that provides a thermal resistance from case to free-air, $R_{\theta CA}$, of not more than 105°C/W.

Input Resistance — This can be given as either a resistance value in ohms or as an input current. In the TL081C the input resistance r_i is given as 10^9 ohms (1 billion ohms!) and the input bias current I_{1B} is given as typically 2 to 3 nanoamperes (2 to 3 billionths of an ampere) in Fig. 13.9.

Power Dissipation — As in all semiconductors, the maximum allowable power dissipation is specified. In Fig. 13.8 this is specified as 670 milliwatts or 0.67 watts for the TL081C - (continuous total dissipation.)

Of course, these are just a few of the specifications for any linear device. Different IC's will have other important characteristics. For example, the comparator switching times and output voltage levels are important. Generally, by looking at the definitions and curves given on these data sheets, you can determine what a given specification means. Each of these specifications will give you some insight into what the device does and how it can be used.

HOW ARE LINEAR IC'S USED?

External resistors set the gain for an operational amplifier.

The most common linear IC is the operational amplifier. It is a general purpose amplifier for performing amplification and other operations. The simplest and most obvious use of the device is as an amplifier. The basic configuration for an inverting amplifier is given in Fig. 13.10. The input is sent to the negative (for inverting) terminal. The output is minus the ratio of the two top resistances times the input. Thus, the gain is determined simply and is held fixed by inexpensive external resistors. For example, if resistor R1

Figure 13-10. Operational Amplifier Configured as Inverting Amplifier

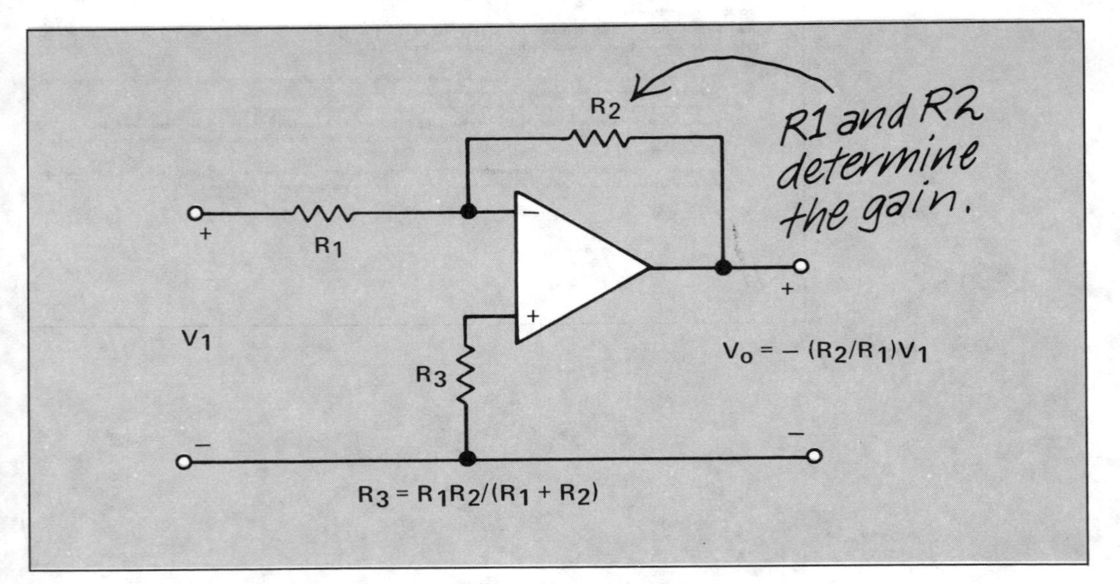

is 10,000 ohms and resistor R2 is 1 million ohms, the output magnitude will be 100 times the input magnitude. In other words, an input of 0.01 volt will cause an output change of 1 volt. If the TL081C is used with these resistors, the out-to-input ratio of 100 will occur over a frequency range of zero to 30,000 hertz. By using such a device, we can build a high performance amplifier in a very small space with an inexpensive IC and three inexpensive resistors. To build such an amplifier with discrete components would require considerable effort, many components, a lot of space, and would be quite expensive.

The values and configuration of the external circuit components determine the type of mathematical operation performed by an operational amplifier.

You can see from this application why these are called amplifiers. But why are they called operational amplifiers? The reason is simple. They can be used to perform mathematical operations such as addition, subtraction, and even integration or logarithmic operations. The simplest of these functions, addition, is illustrated in Fig. 13.11. It uses the same basic configuration as simple amplification, but with more inputs. The output is now the sum of all the input voltages times the gain ratio of the amplifier. Again, the gain ratio is the ratio of the two resistances. Similar simple external component configurations can be used to perform other mathematical operations as required.

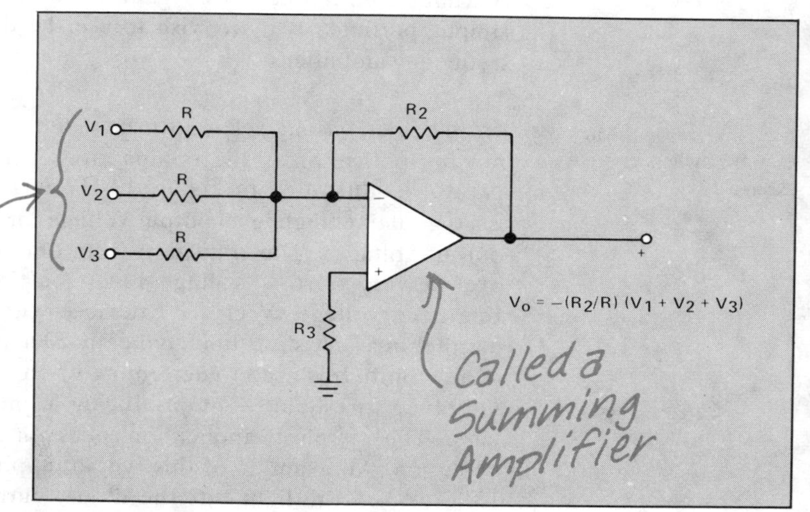

Figure 13-11. Operational Amplifier Configured for Addition

Addition: The sum of these inputs determines the output.

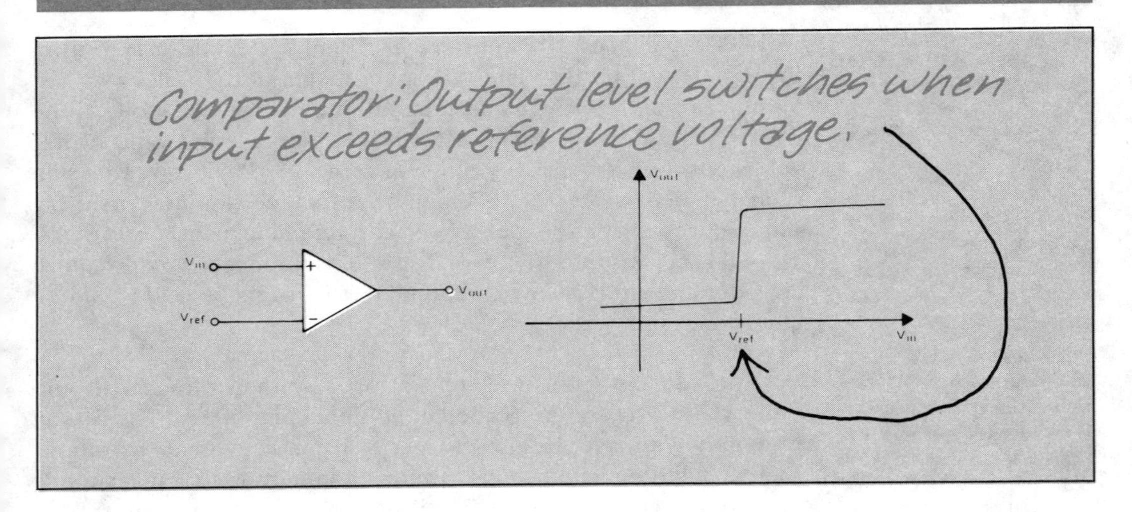

Figure 13-12. Comparator

The operational amplifier is typically a very high gain device but one that may be able to respond to relatively low frequency signals. For amplification of signals with frequencies above a megahertz (1 million hertz) we need a higher frequency amplifier. Video amplifiers and RF amplifiers offer such high frequency operation and can perform such tasks as modulation and tuning out undesired frequencies. Again, the application configuration is simple, particularly if we wish to use the device as a simple high frequency amplifier.

A comparator generates an output when its input exceeds some preset reference threshold.

A device closely related to the video amplifier in circuitry but not in application is the comparator. The purpose of the comparator is illustrated in Fig. 13.12. The output of the comparator is a digital one (high level output voltage) or a digital zero (low level output voltage). The transition from one level to another occurs over a very narrow voltage range at the value given by the reference voltage, Vref. The device switches from one level to another in a very short time, typically 20 nanoseconds (20 billionths of a second). It is used to determine when a voltage goes above the reference threshold — by producing a one output when that occurs. The simplest application of such a device is threshold detection. An example of this type of application is shown in Fig. 13.13. When the light hits the diode, current flows through R1, causing a voltage at the + input. The R2 and R3 resistors define the reference voltage at the − input. When the light reaches the level to get the + terminal voltage to the − terminal voltage, the output switches from zero to a one. This output can be used to turn on a relay, a power transistor, or some other working element.

Figure 13-13. Comparator Configured as Threshold Detector

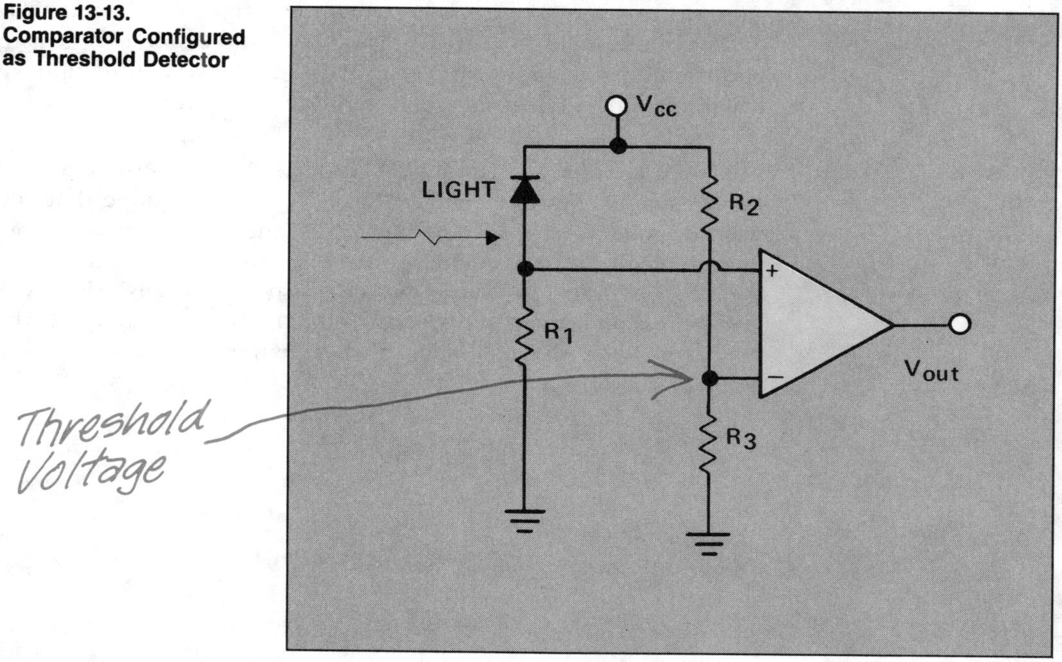

Line drivers amplify digital signals for long-distance transmission. Line receivers recover the digital signals at their destination.

Another device that is closely related to video amplifiers and comparators is the line receiver used in long distance transmission of digital signals. A companion device is the line driver. The two devices are used in the configuration shown in Fig. 13.14. The purpose of these devices is to reliably send very fast digital signals over a distance of several hundred feet even in the presence of large noise voltages. Even though digital signals are transmitted, the devices are considered linear IC's. The line driver converts the input digital signals to current pulses in the transmission line. After traveling over the long distances the current pulses produce very low voltages at the receiver. The receiver must operate with high

Figure 13-14. Line Driver and Line Receiver

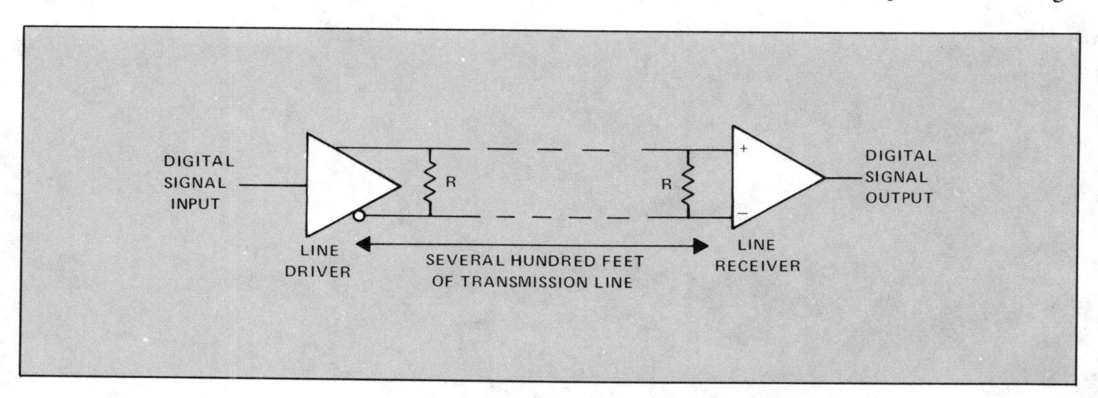

gain and very low thresholds to detect these signals. Again, the use of the circuits is almost obvious. Their interconnection allows us to perform information transmission that would be very difficult if we had to build the driver and receiver devices ourselves.

There are many linear IC's. Naturally, we have just scratched the surface of the types and their applications. We hope that you will continue your study of linear IC's and indeed all semiconductors. You can now build upon the foundation and framework you've established by your study of this book, because the more you understand about semiconductors, circuits, and systems, the better you will enjoy, and help others to enjoy, the benefits of the Semiconductor Age.

Quiz for Chapter 13

1. Most linear ICs are:
   - ☐ a. Very straight in physical appearance — hence the name, "linear"
   - ☐ b. Essentially amplifiers of various kinds
   - ☐ c. Switching type circuits as opposed to digital types
   - ☐ d. Unimportant in the semiconductor business because their use is limited to only a few highly specialized systems.
   - ☐ e. Known to only a few persons

2. Name an important and popular type of linear IC that has extremely high voltage gain and differential inputs, that was originally used for mathematical operations in analog computers but is now used in a very wide variety of applications:
   - ☐ a. Consumer circuits
   - ☐ b. Communications circuits
   - ☐ c. Systems interface circuits
   - ☐ d. Operational amplifiers
   - ☐ e. Sense amplifiers

3. What is the purpose of the group of ICs called "systems interface circuits"?
   - ☐ a. They are unspecialized, general-purpose amplifiers
   - ☐ b. They are used in radio, TV, and automobile electronics
   - ☐ c. They are used mainly in military and industrial radio applications
   - ☐ d. They translate information into and out of analog systems
   - ☐ e. They are input and output units for digital information in digital systems

4. What techniques are used to increase gain in linear IC's?
   - ☐ a. Use of large resistance values
   - ☐ b. Resistor ratio design
   - ☐ c. Use of transistors with wide bases
   - ☐ d. Use of transistors with narrow bases
   - ☐ e. a and c above
   - ☐ f. a and d above

5. Resistor Ratio design is used in linear IC's because:

- ☐ a. Precise resistor values are not possible with IC processes
- ☐ b. Ratios increase amplifier gain
- ☐ c. Ratios increase input resistance
- ☐ d. None of the above
- ☐ e. All of the above except d

6. FET's are used in linear IC's to:

- ☐ a. Increase input resistance
- ☐ b. Increase device complexity
- ☐ c. Provide large resistances
- ☐ d. All of the above
- ☐ e. a and c above

7. A gain of 60 decibels is the same as a gain of:

- ☐ a. 10 volts/volt
- ☐ b. 100 volts/volt
- ☐ c. 1000 volts/volt
- ☐ d. 10000 volts/volt
- ☐ e. None of the above

8. What bandwidth at a gain of 100 volts/volt will an amplifier with a 1 MHz unity gain bandwidth have?

- ☐ a. 1000 hertz
- ☐ b. 10,000 hertz
- ☐ c. 100,000 hertz
- ☐ d. None of the above

9 A comparator:

- ☐ a. Compares digital inputs and provides a linear output
- ☐ b. Compares two input signals and provides a linear output
- ☐ c. Compares two input signals and provides a digital output
- ☐ d. None of the above

10. What is the purpose of the group of ICs called "systems interface circuits"?

- ☐ a. They are unspecialized, general-purpose amplifiers
- ☐ b. They are used in radio, TV, and automobile electronics
- ☐ c. They are used mainly in military and industrial radio applications
- ☐ d. They translate information into and out of analog systems
- ☐ e. They are input and output units for digital information in digital systems

Glossary

Alternating Current: Electrical current, the flow of which reverses direction (or alternates) at regular intervals. Abbreviated ac.

Amplifier: Basically a name given a transistor or circuit that regulates the flow of electrons, as opposed to switching the flow.

Amplitude Modulation: A means whereby information is sent through the circuit by changing (modulating) the amplitude or height of electrical waves.

Analog: Representing physical quantities by regulating circuit current or voltage that represents the quantity.

AND Gate: The output is yes (1) only if all inputs are yes.

Alloy: Method of making PN junctions by melting a metallic dopant so that it dissolves some of the semiconductor material, and then hardens to produce a doped "alloy."

Base (P-region): Area in an NPN transistor from which electrons are withdrawn to make current flow in a circuit.

Bipolar: General name for NPN and PNP transistors, since working current passes through semiconductor material of both polarities (P and N). Also applied to integrated circuits that use bipolar transistors.

Breakdown Voltage: Voltage at which a massive failure to hold back current occurs ($V_{(BR)EBO}$, etc.)

Capacitance: A cause of reactance. The characteristic (of a device or component) that tends to impede changes in voltage by the "capacity" of that device or component to store a charge (electrons).

Capacitor: Electrical device which stores charge and is commonly used to smooth out irregular pulses in electrical current, allowing a more constant flow of electrons.

Clamping: The use of a diode to prevent the voltage in one wire from exceeding the voltage in a second wire.

Code Converter: A decision-making type of digital building block that converts information received at its inputs to another digital code which is transmitted at its outputs. (Also called encoder or decoder.)

Collector (N-region): The N-region of an NPN transistor that collects the emitted electrons and then passes them on through a conductor, completing the electric circuit.

Comparator: A linear circuit that compares two input signals and provides a digital level output depending on the relationship of the input signals. As an example, a "1" for $A>B$ a "0" for $A<B$.

Conductivity: A measure of how easily a semiconductor conducts. For transistors, conductivity is represented by the voltage drop V_{BE} across the base-emitter junction for forward control current, and the voltage drop V_{CE} in the emitter-collector path for saturation working current.

Control Circuit: A low-power circuit used to control the switching or regulating element in a higher-power working circuit.

Counter: A memory-type digital building block that counts pulses received at its input and transmits the cumulative total at its outputs.

Conventional Current: The conventional (or customary) definition of the direction of electric current. It was originally defined by Benjamin Franklin but has since proven to be imaginary. It is imagined to flow from a more positive to a more negative voltage — from a higher to a lower voltage. Actually, nothing flows in that direction. Instead, electrons are flowing from a more negative to a more positive voltage — from a lower to a higher voltage (here, voltage means "conventional" voltage, which is the pressure producing conventional current).

Current: The flow of electrons in coulombs per second. Usually expressed in amperes (amps or A), milliamperes (mA) or microamperes (μA).

Darlington Configuration: An interconnection of transistors, usually used to obtain high input resistance to an amplifier, that reduces the input base current to approximately I_C/B^x, where x is the number of transistors used.

DTL (Diode-transistor logic): Any logic gate circuit that uses several diodes to perform the AND or OR function, followed by one or more transistors to add power to (and possibly invert) the output. Formerly very popular in digital systems but now largely superseded by TTL circuits.

Data Selector: A decision-making type of digital building block that routes data from any one of several inputs to its outputs.

Decibels: A means of expressing the gain of an amplifier in logarithmic measure. The gain of an amplifier G in decibels is:

$$G_{db} = 20 \log_{10} \frac{\Delta V_o}{\Delta V_{in}}$$

20 times the logarithm to the base 10 of the voltage gain is the gain in decibels.

Detection: The "demodulation" of high-frequency electrical waves from a radio receiving antenna to recover the information (typically sound) being transmitted. This can be done simply (for AM waves) by rectifying the antenna current with a diode.

Diffused Junction: Method of producing PN junctions by exposing masked area of heated semiconductor material to a gaseous dopant (P or N), which soaks in to certain depths.

Digital Integrated Circuit: A switching-type integrated circuit.

Digital: Method of sending information through an electrical circuit by switching the current on or off.

Direct Current: Flow of electrons which only goes in one direction. Abbreviated dc.

Drain: The working-current terminal (at one end of the channel in an FET) that is the drain for holes or free electrons from the channel. Corresponds to collector of bipolar transistor.

ECL (Emitter-Coupled Logic): a type of circuit used as a logic gate that is notable for very fast operating speed and high power dissipation. Used mainly in large, very-high-speed digital computers and sold mainly on a custom-designed basis.

Electric Charge: The property of electrons and protons that makes "like particles repel and unlike particles attract."

Electron: Electrically charged particle orbiting the nucleus of every atom. The charge of the electron is called negative (−) to signify an electrical effect equal and opposite to that of a proton.

Electrons: Tiny particles making up electricity.

Emitter (N-region): The region in an NPN transistor that emits a relatively large number of electrons as a relatively small number of electrons is extracted from the P-region base.

Epitaxial: Means of growing a layer of monocrystalline silicon from silicon vapor. The vapor usually contains dopant gas to produce a PN junction, or to produce a more lightly doped layer of the same type as the substrate.

Fan-Out: An electrical specification of a digital building block. The number of other building blocks that the first block is capable of transmitting to. The number of inputs that one output can drive, determined by the current capability of the output divided by the current requirement of an input.

FET (Field-Effect Transistor): A transistor controlled by voltage rather than current. The flow of working current through a semiconductor channel is switched and regulated by the effect of an electric field exerted by electric charge in a region close to the channel called the gate; also called unipolar transistor. An FET has either P-channel or N-channel construction.

Flip-Flop: A digital building block that, upon command from a "clock pulse" received at one input, stores (that is, holds or remembers) at its output a bit of information (logical 1 or 0) received at another input.

Forward Current (I_F): The amount of conventional current flowing from anode to cathode when a given forward voltage is imposed on a diode.

Forward Voltage (V_F): In a diode, the amount by which anode voltage exceeds (is more positive than) cathode voltage. (Voltage here refers to conventional or positive voltage — the imaginary pressure of positive charge.)

Frequency: How many times every second an alternating current goes through a complete cycle (turning around backwards, and then going forward again). Formerly expressed in cycles per second (cps) and multiples. Now expressed in equivalent units of hertz (Hz), kilohertz (kHz), megahertz (MHz) and gigahertz (GHz).

Frequency Modulation: A means whereby analog information is sent through the circuit by changing (modulating) the frequency of the electrical waves.

Frequency Response: The gain of an amplifier plotted against the frequency of the signal being amplified.

Gain: Ratio of output signal (working circuit) to input signal (control circuit). For an NPN or PNP transistor, the type of gain specified in data sheets is current gain, symbolized h_{fe} or h_{FE}, and often called "beta".

Gate: The control terminal and controlling region of a field-effect transistor. Corresponds to base of bipolar transistor.

Half-Adder: An example of a building block used in digital computers. It is a simple combination of logic gates that adds two bits. The answer is two bits, called sum and carry. Enough half-adders combined in a larger building block can add two binary numbers of any length.

Inductance: A cause of reactance. The characteristic (of a device or component) that tends to impede changes in current passing through the device or component, as if the electrons had inertia or sluggishness.

Industrial-Grade IC: Typically, an IC whose performance is guaranteed over the temperature range from 0°C to 70°C.

Integrated Circuit (IC): An electronic circuit — containing transistors diodes, resistors, and perhaps capacitors, along with interconnecting electrical conductors — processed and contained entirely within a single chip of silicon.

Junction-FET: An FET whose gate element is a region of semiconductor material (ordinarily, the substrate) insulated by a PN junction from the channel, which is material of opposite polarity. All junction-FETs are depletion-type (normally turned on).

Leakage Current: Unwanted current flowing in a semiconductor where it ought to be blocked. (I_{EBO},I_{CBO}, etc.)

Linear Integrated Circuit: An amplifying-type integrated circuit whose output varies over a continuous range rather than just discrete levels, as in digital.

Logarithm (of a number): The power that the base must be raised to in order to get the number.

Logic Gates: Switching-circuit building blocks which utilize yes and no statements as inputs to make certain simple decisions with the answer also expressed as yes or no. The yes and no are represented as voltage or current levels.

Maximum Power Dissipation: Semiconductor specification relating to the amount of heat the device can stand before it malfunctions. This is sometimes expressed in terms of current: as "maximum forward current" I_F for diodes, and "maximum collector current" I_C for transistors.

Mesa Diffusion: Diffused junction method of growing PN junctions where a single base region is created over the entire slice. Acid is used to etch away valleys between emitters, leaving "mesas" of processed material for use as transistor elements.

Microcomputer: A computer in the lowest range of size and speed which contains a CPU, stored instructions in ROM memory, RAM memory for data and instructions and I/O circuits.

Microprocessor: An IC (or set of ICs) that can be programmed with stored instructions to perform a variety of functions. It has the basic parts of a simple CPU — control, timing, ALU.

Military-Grade IC: Typically, an IC whose performance is guaranteed over the temperature range from −55°C to +125°C.

Modulator: An amplifying-type circuit function whose output is a copy of oscillating electrical waves at its input, except that some characteristic, such as the amplitude (height), of the output waves is modulated (controlled) by a second input. (This is an amplitude modulator, and is the most important circuit function in an AM radio transmitter.)

Monocrystalline: Made up of a single, continuous crystal. This is the form required for semiconductor material used in electronic devices.

MOS: Metal-oxide-semiconductor, referring to a field-effect transistor (MOSFET) that has a metal gate insulated by an oxide layer from the semiconductor channel. A MOSFET is either enhancement-type (normally turned off) or depletion-type (normally turned on). MOS also refers to integrated circuits that use MOSFETs (virtually all enhancement-type).

Multiplexer: A decision-making type of digital building block that routes data from its one input to any one of several outputs.

NAND Gate: NAND (NOT-AND) gate is an AND gate followed by an inverter. The output of the AND gate is inverted to the opposite value.

Noise Margin: An electrical specification of a digital building block that describes how securely the block transmits and receives correct information in spite of random, unwanted electrical signals called "noise" that tend to creep into any communications link. It is the "safety margin" between the output voltages V_{OH} and V_{OL} produced by the transmitting block and the input voltages V_{IH} and V_{IL} required by the receiving block.

NOR Gate: NOR (NOT-OR) is an OR gate followed by an inverter. The output of the OR gate is inverted to the opposite value.

NOT Gate: The output is just the opposite from the single input.

Negative Logic: Letting the most negative of the two logic voltages stand for "one" or "yes," and the least negative voltage stand for "zero" or "no."

Noise Figure: A measure of how much unwanted signal is generated in the process of amplification, "noise" being the result of turbulent, uneven rushing of electrons.

N-type Semiconductor Material: A semiconductor crystal containing a small proportion of "dopant" atoms that have one more outer electron than the other atoms. These negative extra electrons can find no unoccupied orbit to bind them, so they are free to wander and constitute electric current. A common N-type dopant for silicon is phosphorus.

Op-amp: Operational amplifier. An unspecialized amplifier very commonly used in IC form, characterized by extremely high voltage gain and differential inputs. Originally used for mathematical operations in analog computers, but now employed virtually anywhere a low-power amplifier is needed.

Operating Speed: Electrical (sometimes called "switching") specifications of a digital building block that describe how fast the block operates ("makes a decision"). Usually expressed as propagation delay times t_{PLH} and t_{PHL}.

OR Gate: The output is yes (1) if any or all of its inputs are yes (1).

Oscillator: An amplifying-type circuit function whose output is a regularly fluctuating (oscillating) current or voltage.

P-N Junction: The dividing line in a semiconductor between a P region and an N region. Electrons can flow from N to P but not from P to N.

P-N Junction Diode: Semiconductor device which permits electrons to flow through it in one direction only.

Phase Control: A method of regulating a supply of alternating current by use of a switching device such as a thyristor, by varying the point in each
a-c cycle or half-cycle at which the device is switched on.

Photodiode: A PN semiconductor diode designed so that light falling on it greatly increases the reverse leakage current, so that the device can switch and regulate electric current in response to varying intensity of light.

Photomask: A transparent plate slightly larger than a silicon slice, containing numerous tiny opaque spots, used in the planar diffusion process as a shadowmask over a slice coated with photoresist to expose the surface of the slice to acid in desired spots in a later step.

Photoresist: A liquid plastic which hardens into a tough, acid-resistant solid when exposed to ultraviolet light.

Phototransistor: A transistor whose base-collector junction acts as a photodiode, so that light generates a base current which turns on the working current through the transistor. This gives a much larger current than a simple photodiode.

Planar Diffusion: Diffused junction method of growing PN junctions where all junctions emerge at the flat upper surface of the slice.

PNP: Identical in principle and application to the NPN transistor, except that the positions and functions of P-type material and N-type material are reversed, along with the functions of holes and free electrons. As a result, current is also reversed, so a PNP is useful where a transistor is needed to turn on when conventional current is withdrawn from the base.

Polycrystalline: Made up of numerous tiny crystals joined together in disorderly fashion. This is the form in which silicon or germanium emerges from the chemical purification step.

Positive Logic: Letting the most positive of the two logic voltages stand for "one" or "yes", and least positive voltage stand for "zero" or "no."

Power Dissipation: An electrical specification of a digital building block (usually not formally guaranteed on a data sheet because it depends on so many variables) that describes how many watts of power are consumed by the block and converted to heat during normal operation.

Power Dissipation: Whenever electric current flows from a higher to a lower voltage — such as through a motor or resistor or transistor — a certain amount of energy has to come out of it (measured in watts: equal to amps multipled by voltage drop). If this energy or power is not converted to work (for example, by a motor), it is "dissipated" — that is, wasted in the form of heat.

P-type Semiconductor Material: A semiconductor crystal containing a small proportion of "dopant" atoms that have one less outer electron than the other atoms. Each dopant atom causes one unoccupied spot — a "hole" — among the electrons that are bound in their orbits. The holes are positively charged and can move to constitute electric current. Boron is a commonly used P-type dopant for silicon.

Proton: Electrically charged particle within the nucleus of every atom. The charge of the proton is called positive (+) to signify that it is equal to and opposite from that of the electron.

RAM (Random-Access Memory): A certain kind of memory-type building block commonly available in MOS IC form, some consist of flip-flops for cells, other just contain a capacitor with gating circuits for reading, writing and refreshing. Bits of binary information can be "written" into or "read" out of any of the cells at will.

Reactance: The property of an electrical device or conductor that tends to impede changes in ac current passing through it or ac voltage exerted on it. ("Reactance" determines the way a circuit "reacts" to changes in ac current or ac voltage.)

Read-Only Memory (ROM): A digital building block usually classified as memory-type that contains information permanently written into it during manufacture which can be "read" at the outputs, but not changed ("written").

Rectification: The most straightforward use of a diode — the conversion of alternating current to pulses of direct current.

Register: A group of flip-flops used together to store several bits. One variety is the shift register.

Resistance: Difficulty in moving electrical current through a circuit when voltage is applied. Usually expressed in ohms (Ω) or kilohms (kΩ).

Reverse Breakdown Voltage $V_{(BR)}$: The reverse voltage beyond which a diode cannot hold back reverse current. Effectively the junction breaks down and can be damaged in this condition.

Reverse Current (I_R): The amount of conventional current flowing from cathode to anode when a given reverse voltage is imposed on a diode.

Reverse Recovery Time(t_{rr}): The time it takes for the diode to recover from forward conduction and begin to block reverse current.

Reverse Voltage (V_R): In a diode, the amount by which the cathode voltage exceeds the anode voltage.

SCR (Silicon Controlled Rectifier): Formal name is "reverse-blocking triode thyristor." A thyristor that can be triggered into conduction in only one direction. Terminals are called "anode," "cathode," and "gate."

Semiconductor Material: A crystal (usually silicon or germanium) that is "sometimes" a conductor because it can be made to carry electric current by means of free electrons, or by means of "holes" among bound electrons.

Shift Register: A digital building block consisting of any number of flip-flops connected in series with a common clock signal, so that with a clock pulse each bit of information being stored shifts to the next flip-flop in the chain; one bit at a time, all at the same time.

Source: The working-current terminal (at one end of the channel in an FET) that is the source for holes (P-channel) or free electrons (N-channel) flowing in the channel. Corresponds to emitter of bipolar transistors.

SSI, MSI, LSI, VLSI: A level of complexity of integrated circuits allowing the fabrication of circuitry with respectively, up to 12 gates (SSI), up to 100 gates (MSI), up to 1000 gates (LSI) and complete systems with over 50,000 gates (VLSI).

Systems Interface Circuits: Name used at Texas Instruments for a group of ICs that act as input and output units ("interfaces") for information in digital systems. This general function is basically a matter of amplifying digital signals. Included are memory drivers, sense amplifiers, line drivers and receivers, and peripheral drivers.

TTL (Transistor-Transistor Logic): Any logic gate circuit that uses a multiemitter NPN transistor to perform the positive AND function, followed by one or more transistors to add power to (and possibly invert) the output. Operating speed is generally faster than that of DTL.

TTL 54-74 Series: A series (originated by Texas Instruments) of digital building blocks that use as the basic gate unit one of a group of certain very similar TTL circuits.

Triac: Formal name is "bidirectional triode thyristor." A thyristor that can be triggered into conduction in either direction. Terminals are called "main terminal 2," "main terminal 1," and "gate."

Thyristor: One of a class of semiconductor devices (made of at least four alternate P-N-P-N layers) that snap to a completely "on" state for working current when a momentary pulse of control current is received, and (typically) can be turned off only by interrupting the working current elsewhere in the circuit. Thyristors are mostly high-power devices.

Voltage Gain: For an amplifier, the ratio of the change in output voltage to the change in input voltage, usually express as a non-dimensional number but can be expressed in decibels.

Voltage: Electron pressure in an electrical wire or circuit. Usually expressed in volts (V).

Working Circuit: A circuit that provides electrical power to a device that does work or transmits information.

Index

Quiz Answers

Chapter 1

1. **c**
2. **b**
3. **d**
4. **a**
5. **d**
6. **c**
7. **c**
8. **b**
9. **a**
10. **c**

Chapter 2

1. **a**
2. **c**
3. **b**
4. **c**
5. **c**
6. **c**
7. **a**
8. **b**
9. **d**
10. **d**
11. **b**
12. **d**

Chapter 3

1. **e**
2. **c**
3. **b**
4. **c**
5. **b**
6. **b**
7. **e**
8. **e**
9. **e**
10. **e**

Chapter 4

1. **e**
2. **b**
3. **a**
4. **b**
5. **e**
6. **d**
7. **e**
8. **d**
9. **f**
10. **g**

Chapter 5

1. **e**
2. **a**
3. **a**
4. **a**
5. **e**
6. **e**
7. **a**
8. **d**
9. **a**
10. **e**
11. **e**

Chapter 6

1. **d**
2. **a**
3. **d**
4. **a**
5. **d**
6. **e**
7. **e**
8. **a**
9. **a**
10. **a**
11. **b**

Chapter 7

1. **b**
2. **a**
3. **d**
4. **a**
5. **c**
6. **b**
7. **c**
8. **c**
9. **d**
10. **e**

Chapter 8

1. **a**
2. **c**
3. **d**
4. **e**
5. **b**
6. **e**
7. **e**
8. **a**
9. **b**
10. **c**

Chapter 9

1. **e**
2. **d**
3. **c**
4. **a**
5. **b**
6. **a**
7. **e**
8. **c**
9. **e**
10. **e**

Chapter 10

1. **c**
2. **e**
3. **e**
4. **d**
5. **a**
6. **b**
7. **d**
8. **c**
9. **e**
10. **a**

Chapter 11

1. **b**
2. **e**
3. **c**
4. **a**
5. **a**
6. **b**
7. **d**
8. **d**
9. **b**
10. **e**

Chapter 12

1. **c**
2. **e**
3. **d**
4. **c**
5. **a**
6. **f**
7. **f**
8. **e**
9. **e**

Chapter 13

1. **b**
2. **d**
3. **e**
4. **f**
5. **a**
6. **e**
7. **c**
8. **b**
9. **c**
10. **e**